LOGIC AND
BOOLEAN
ALGEBRA

BY

KATHLEEN LEVITZ

ASSISTANT UNITED STATES ATTORNEY

HILBERT LEVITZ

FLORIDA STATE UNIVERSITY

BARRON'S EDUCATIONAL SERIES, INC.

WOODBURY, NEW YORK

All inquiries should be addressed to:
Barron's Educational Series, Inc.
113 Crossways Park Drive
Woodbury, New York 11797

Library of Congress Catalog No. 75-1006
International Standard Book No. 0-8120-0537-6

Library of Congress Cataloging in Publication Data
Levitz, Kathleen.
Logic and Boolean algebra.
Bibliography:
Includes index.
1. Logic, Symbolic and mathematical. 2. Algebra, Boolean. I. Levitz, Hilbert, joint author. II. Title.
QA9.L47 511'.3 75-1006
ISBN 0-8120-0537-6

PRINTED IN THE UNITED STATES OF AMERICA

TO OUR MENTORS

JOE L. MOTT
KURT SCHÜTTE

Contents

Preface

This book was intended for students who plan to study in the humanities and in the social and management sciences. Students interested in the physical and natural sciences, however, might also find its study rewarding. All that is presupposed is some high school algebra. The authors strongly urge that the topics be studied in the order in which they appear and that no topics be skipped.

In recent years there has been considerable divergence of opinion among mathematics teachers as to the degree of abstraction, rigor, formality, and generality that is appropriate for elementary courses. We believe that the trend lately has been to go too far in these directions. Quite naturally, this book reflects our views on this question. Although the subject matter is considered from a modern point of view, we have consciously tried to emulate the informal and lucid style of the better writers of a generation ago. Manipulative skills are cultivated slowly, and the progression from the concrete to the abstract is very gradual.

We wish to extend our thanks to our typist Susan Schreck and to Matthew Marlin and Anne Park, who were students in a course from which this book evolved. We also wish to thank the editorial consultants of Barron's for their helpful suggestions.

Tallahassee, Florida

KATHLEEN LEVITZ
HILBERT LEVITZ

Introduction

Logic is concerned with reasoning. Its central concern is to distinguish good arguments from poor ones. One of the first persons to set down some rules of reasoning was Aristotle, the esteemed philosopher of ancient Greece. For almost two thousand years logic remained basically as Aristotle had left it. Students were required to memorize and recite his rules, and generally they accepted these rules on his authority.

At the end of the eighteenth century Kant, one of the great philosophers of modern times, expressed the opinion that logic was a completed subject. Just fifty years later, however, new insights and results on logic started to come forth as a result of the investigations of George Boole and others. In his work, Boole employed symbolism in a manner suggestive of the symbolic manipulations in algebra. Since then, logic and mathematics have interacted to the point that it no longer seems possible to draw a boundary line between the two.

During the last forty years some deep and astounding results about logic have been discovered by the logician Kurt Gödel and others. Unfortunately, these results are too advanced to be presented in this book. We hope that what you learn here will stimulate you to study these exciting results later.

Finally, we must tell you that complete agreement does not yet exist on the question of what constitutes correct reasoning. Even in mathematics, where logic plays a fundamental role, some thoughtful people disagree on the correctness of certain types of argumentation. Perhaps someday, someone will settle these disagreements once and for all.

1

Sentence Composition

1.1
The Basic Logical Operations

Compound sentences are often formed from simpler sentences by means of the five **basic logical operations.** These operations and their symbols are:

conjunction $\wedge$
disjunction $\vee$
negation $\neg$
implication $\rightarrow$
bi-implication $\leftrightarrow$

The symbols are usually read as follows:

SYMBOL	TRANSLATION
$\wedge$	**and**
$\vee$	**or**
$\neg$	**not**
$\rightarrow$	**if . . . then . . .**
$\leftrightarrow$	**if and only if**

Note that a good way to keep from confusing the symbols $\wedge$ and $\vee$ is to remember that $\wedge$ looks like the first letter of the word "AND."

If the letters A and B denote particular sentences, you can use the logical operations to form these compound sentences:

$A \wedge B$, read A *and* B
$A \vee B$, read A *or* B
$\neg A$, read *it is not the case that* A
$A \rightarrow B$, read *if* A *then* B
$A \leftrightarrow B$, read A *if and only if* B

Examples. Let A be the sentence: "Snow is white."
Let B be the sentence: "Grass is green."
Then $A \wedge B$ is the sentence: "Snow is white and grass is green."

Let A be the sentence: "Humpty Dumpty is an egg."
Then $\neg A$ is the sentence: "It is not the case that Humpty Dumpty is an egg."

Let A be: "Jack is a boy."
Let B be: "Jill is a girl."
Then $A \rightarrow B$ is: "If Jack is a boy, then Jill is a girl."

Let A be: "Birds fly."
Let B be: "Bees sting."
Let C be: "Bells ring."
Then $\neg A \rightarrow (B \vee C)$ is: "If it is not the case that birds fly, then bees sting or bells ring."

Note that in the last example, the "not" sign applies only to the sentence A. If we had wanted to negate the entire sentence $A \rightarrow (B \vee C)$, we would have enclosed it in parentheses and written $\neg [A \rightarrow (B \vee C)]$. Then it would read, "It is not the case that if birds fly, then bees sting or bells ring."

1.2
Truth Values

A property of declarative sentences is that they are true or false, but not both. If a sentence is true, it has **truth value t.** If it is false, it has **truth value f.** You can compute the truth value of a compound sentence built from simpler sentences and logical operations if you know the truth values of the simpler sentences. This can be done by means of tables. The table for the conjunction operation is given below. Here is how to read it. On a given row, the extreme right-hand entry shows the truth value that the compound sentence $A \wedge B$ should have if the sentences A and B have the truth values entered for them in that row.

TABLE FOR $\wedge$ ("AND")

A	B	$A \wedge B$
t	t	t
t	f	f
f	t	f
f	f	f

Examples. Use the table just given to find the truth values of these compound sentences:

a) Giants are small and New York is large.
b) New York is large and giants are small.
c) America is large and Russia is large.

ANSWERS: a) First label each part of the sentence with its own truth value.

$$\overbrace{\text{[Giants are small]}}^{\text{f}} \wedge \overbrace{\text{[New York is large]}}^{\text{t}}.$$

Now look in the table to see which row has the values f, t (in that order) entered in the two left-hand columns. This turns out to be the third row (below the headings). Looking to the extreme right of that row, you will find that the entire sentence has the truth value f.

b) Labeling each part with its truth value, we get

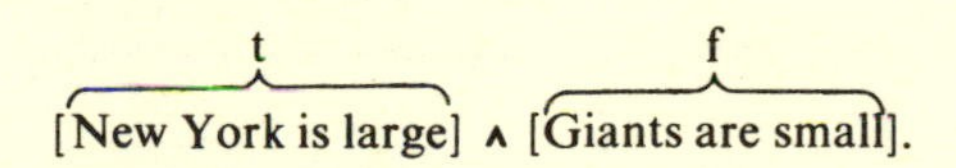

Now look in the table to see which row has the values t, f (in that order) entered in the two left-hand columns. This turns out to be the second row. Looking to the extreme right of that row, you will find that the entire sentence has the truth value f.

c) First label the parts:

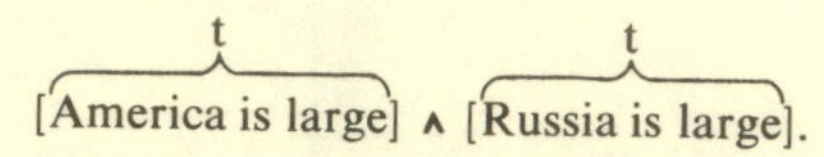

The first row of the table indicates that the entire sentence has the truth value t.

We make the tables for the other four logical operations in a similar way:

TABLE FOR ∨ ("OR")

A	B	A ∨ B
t	t	t
t	f	t
f	t	t
f	f	f

TABLE FOR ¬ ("NOT")

A	¬ A
t	f
f	t

TABLE FOR → ("IF ... THEN ...")

A	B	A → B
t	t	t
t	f	f
f	t	t
f	f	t

TABLE FOR ↔ ("IF AND ONLY IF")

A	B	A ↔ B
t	t	t
t	f	f
f	t	f
f	f	t

According to the table for ∨, the disjunction A ∨ B is a true sentence if A is true, if B is true, or if both A and B are true. Unfortunately, in ordinary conversation people do not always use "or" this way. However, in mathematics and science (and in this book), the sentence A ∨ B is considered true even in the case where A and B are both true.

The implication operation presents a similar problem. Quite often "if A, then B" indicates a cause-and-effect relationship as in the sentence:

If it rains, the game will have to be postponed.

Mathematicians and scientists, however, do not require such a cause-and-effect relationship in affirming the truth of A → B, and our table has been set up in accordance with their time-honored conventions.

Examples. Use the tables just given to compute the truth values of the following sentences:

a) If two equals six, then three equals three.
b) $(5 = 7) \vee (6 = 8)$.
c) It is not the case that three equals three.
d) $[(2 = 4) \vee (3 = 3)] \rightarrow [(5 = 0) \wedge (6 = 1)]$.

ANSWERS: First label the parts with their truth values:

a) $\overbrace{\text{[two equals six]}}^{f} \rightarrow \overbrace{\text{[three equals three]}}^{t}$

The third row of the table for $\rightarrow$ shows that the entire sentence gets the truth value t.

b) $\overbrace{(5 = 7)}^{f} \vee \overbrace{(6 = 8)}^{f}$

The fourth row of the table for $\vee$ shows that the entire sentence gets the truth value f.

c) $\neg\ \overbrace{\text{[three equals three]}}^{t}$

The first row of the table for $\neg$ shows that the entire sentence gets the truth value f.

d) This one will involve the use of three tables because it contains three logical operation symbols.
First label the elementary parts:

$$[\overbrace{(2 = 4)}^{f} \vee \overbrace{(3 = 3)}^{t}] \rightarrow [\overbrace{(5 = 0)}^{f} \wedge \overbrace{(6 = 1)}^{f}].$$

From the third row of the table for $\vee$ you can see that the part to the left of the arrow gets the truth value t. From the fourth row of the table for $\wedge$, you can see that the part to the right of the arrow gets the truth value f. Now labeling the parts to the left and right of the arrow with the truth values just computed for them, you have

$$\overbrace{[(2 = 4) \vee (3 = 3)]}^{t} \rightarrow \overbrace{[(5 = 0) \wedge (6 = 1)]}^{f}.$$

From the second row of the table for → you can see that the entire sentence should have the truth value f.

Exercises 1.2

1. Let A denote "Geeks are foobles" and let B denote "Dobbies are tootles." Write the English sentences corresponding to the following:

(a) $\neg A$
(b) $A \wedge B$
(c) $A \wedge \neg B$
(d) $\neg A \rightarrow B$
(e) $\neg (A \rightarrow B)$
(f) $A \leftrightarrow B$
(g) $(\neg A \vee B) \rightarrow A$
(h) $(A \wedge B) \leftrightarrow \neg B$

2. Let A denote "Linus is a dog," let B denote "Linus barks," and let C denote "Linus has four legs." Write each of the following sentences in symbolic form:

(a) Linus is a dog and Linus barks.
(b) Linus is a dog if and only if Linus barks.
(c) If it is not the case that Linus is a dog, then Linus barks.
(d) If Linus is a dog, then (Linus barks or Linus has four legs).
(e) If (Linus barks and Linus has four legs), then Linus is a dog.
(f) (It is not the case that Linus barks) if and only if Linus is a dog.
(g) It is not the case that (Linus is a dog if and only if Linus barks).

3. Let A denote "$1 + 1 = 2$" and let B denote "$2 \cdot 2 = 2$." Use the tables to find the truth values of the following sentences:

(a) $A \vee B$
(b) $A \vee \neg B$
(c) $\neg A \vee B$
(d) $\neg A \wedge B$
(e) $A \wedge \neg B$
(f) $\neg A \rightarrow B$
(g) $\neg A \rightarrow \neg B$
(h) $B \rightarrow A$
(i) $\neg A \rightarrow (\neg B \wedge A)$
(j) $\neg A \leftrightarrow B$
(k) $\neg B \leftrightarrow \neg A$
(l) $(\neg A \vee B) \vee (A \wedge B)$
(m) $(B \vee A) \vee \neg (B \vee A)$
(n) $(B \rightarrow A) \rightarrow (A \rightarrow B)$

4. Let A and B be sentences. Assuming that B has truth value f, use the tables to find those truth values for A which make the following sentences true.

(a) $B \rightarrow A$
(b) $A \wedge B$
(c) $\neg A \leftrightarrow B$
(d) $(A \vee B) \rightarrow B$
(e) $(\neg B \rightarrow A) \rightarrow A$

1.3 Alternative Translations

In English there are many ways of saying the same thing. Here is a list of some of the alternative ways which can be used to translate the logical operation symbols.

$A \wedge B$	**A and B** **A but B** **Both A and B**	**Not only A but B** **A although B**
$A \vee B$	**A or B** **A or B or both**	**Either A or B** **A and/or B [found in legal documents]**
$\neg A$	**A doesn't hold** **It is not the case that A**	
$A \rightarrow B$	**If A, then B.** **A only if B** **A implies B** **B provided that A**	**A is a sufficient condition for B** **B is a necessary condition for A**
$A \leftrightarrow B$	**A if and only if B** **A exactly when B**	

Exercises 1.3

1. Let A be: Peter is a canary.
   Let B be: Joe is a parakeet.
   Let C be: Peter sings.
   Let D be: Joe sings.

Translate the following into symbols.

(a) Joe is a parakeet and Peter is a canary.
(b) Although Joe does not sing, Peter sings.
(c) Peter sings if and only if Joe does not sing.
(d) Either Peter is a canary or Joe is a parakeet.
(e) Peter sings only if Joe sings.

2. Let M be: Mickey is a rodent.
   Let J be: Jerry is a rodent.
   Let T be: Tom is a cat.

Translate the following into symbols.

(a) Although Mickey is a rodent, Jerry is a rodent also.
(b) Mickey and Jerry are rodents, but Tom is a cat.
(c) If either Mickey or Jerry are rodents, then Tom is a cat.
(d) Jerry is a rodent provided that Mickey is.
(e) Either Mickey isn't a rodent or Jerry is a rodent.
(f) Jerry is a rodent only if Mickey is.
(g) Tom is a cat only if Jerry isn't a rodent.

1.4
Converses and Contrapositives

Suppose you are given an implication $A \rightarrow B$. Two related implications are given special names.

$B \rightarrow A$ is called the **converse** of $A \rightarrow B$.

$\neg B \rightarrow \neg A$ is called the **contrapositive** of $A \rightarrow B$.

The truth value of an implication and its converse may or may not agree. Below are given some examples to show this. Later you will see that *an implication and its contrapositive always have the same truth value.*

Examples. Let A be: $1 = 2$

Let B be: 2 is an even number.

Then $A \rightarrow B$ has truth value t, while its converse $B \rightarrow A$ has truth value f.

Let E be: 2 is an even number.

Let 0 be: 3 is an odd number.

Then $E \rightarrow 0$ has truth value t, and its converse $0 \rightarrow E$ also has truth value t.

Exercises 1.4

1. Let D be: Ollie is a dragon.
Let T be: Ollie is toothless.
Let R be: Ollie roars.

Represent each of the following sentences symbolically. Then write the converse and the contrapositive of each sentence, both in symbols and in English.

(a) If Ollie is a dragon, then Ollie roars.
(b) If Ollie is toothless, then Ollie does not roar.
(c) Ollie roars only if Ollie is a dragon.

2. Let A and B be two sentences. If A has truth value t, which truth value must B have to insure that:

(a) $A \rightarrow B$ has truth value t?
(b) the contrapositive of $A \rightarrow B$ has truth value t?
(c) the converse of $A \rightarrow B$ has truth value t?

3. Give examples of implications that have truth value t such that:

(a) the converse has truth value t.
(b) the converse has truth value f.

1.5
Logic Forms and Truth Tables

Logical symbolism enables us to see at a glance how compound sentences are built from simpler sentences. Meaningful expressions built from variable symbols, logical operation symbols, and parentheses are called **logic forms.** Capital letters like A, B, C ... can be used as the variable symbols in constructing logic forms. The following are examples of logic forms:

$$A \rightarrow B$$
$$B$$
$$A \vee C$$
$$(\neg A \rightarrow B) \rightarrow B$$

Note that a single variable symbol is acceptable as a logic form.

Logic forms will often be discussed. When talking *about* logic forms, heavy-type capital leters like **A, B, C** (with or without numerical subscripts) will be used to denote them. Thus in a given discussion, you might use the symbol **B** to denote the logic form

$$(B \rightarrow A) \rightarrow (C \rightarrow A)$$

Remember:

1. Heavy type capital letters, like **A**, denote entire logic forms.
2. Regular capital letters, like A, are variable symbols which appear in a logic form.

You have already seen that the truth value of a logic form is determined by the truth values assigned to its variable symbols. Let A and B be variable symbols. A preceding section contained tables listing the values assigned to each of the logic forms:

$$\neg A \qquad A \vee B \qquad A \wedge B \qquad A \rightarrow B \qquad A \leftrightarrow B$$

for given truth values of A and B. The tables are examples of **truth tables.** Using these five tables, you can construct the truth table of *any* logic form. As an example, consider the truth table of the logic form

$$(A \vee C) \rightarrow (B \rightarrow A).$$

In each row give an assignment of truth values to the variable symbols A, B, and C, and at the right-hand end of the row list the value of the entire logic form for that assignment. The optional columns headed "A v C" and "B → A" are included to help fill in the table. If you feel sufficiently confident, you can omit such intermediate columns when building future truth tables.

TRUTH TABLE FOR (A ∨ C) → (B → A)

A	B	C	A ∨ C	B → A	(A ∨ C) → (B → A)
t	t	t	t	t	t
t	t	f	t	t	t
t	f	t	t	t	t
t	f	f	t	t	t
f	t	t	t	f	f
f	t	f	f	f	t
f	f	t	t	t	t
f	f	f	f	t	t

optional columns

We suggest that you first fill in the columns under the variable symbols A, B, C . . . and then complete these columns one at a time. Below are repeated the first three columns of the preceding table. Note that the places in the columns which have t as entries are shaded, and observe that there is a pattern to this shading. By following the pattern you can be sure that a logic form with n distinct variable symbols will have a truth table with 2^n rows and that these rows display *all* the possible ways of assigning truth values to the variable symbols. Note that the top half of the left-hand column consists only of t entries, the bottom half of f entries. In the next column blocks of t entries and blocks of f entries are alternated; each block consists of 1/4 the total number of rows in the table. In the third column, each block makes up 1/8 of the total number of rows, etc.

A	B	C
t	t	t
t	t	f
t	f	t
t	f	f
f	t	t
f	t	f
f	f	t
f	f	f

Exercises 1.5

1. Construct the truth table for each of the following logic forms:

(a) $A \rightarrow \neg A$ (b) $(C \wedge \neg B) \vee (A \rightarrow B)$
(c) $\neg (A \vee C)$ (d) $[(A \rightarrow B) \vee (C \rightarrow \neg D)] \rightarrow (A \wedge \neg D)$
(e) $[(B \vee A) \wedge C] \leftrightarrow [(B \wedge C) \vee (A \wedge C)]$

2. Use truth tables to determine the truth value of D, given the following:

(a) C is true and $C \wedge D$ is true
(b) C is false and $D \vee C$ is true
(c) $\neg D \vee \neg C$ is false
(d) C is false and $(C \wedge D) \rightarrow (C \vee D)$ is true

3. a) Write a logic form corresponding to the following sentence:
If the stock's value rises or a dividend is declared, then the stockholders will meet if and only if the board of directors summons them, but the chairman of the board does not resign.

b) Determine the truth value of the preceding statement under the following assumptions by (partially) filling out the truth table for its logic form.

i) The stock's value rises, no dividend is declared, the stockholders will meet, the board of directors summons the stockholders, and the chairman resigns.

ii) The stock's value rises and a dividend is declared, the board of directors fails to summon the stockholders, and the chairman resigns.

4. Horace, Gladstone, and Klunker are suspected of embezzling company funds. They are questioned by the police and testify as follows:
Horace: Gladstone is guilty and Klunker is innocent.
Gladstone: If Horace is guilty, then so is Klunker.
Klunker: I'm innocent, but at least one of the others is guilty.

(a) Assuming everyone is innocent, who lied?

(b) Assuming everyone told the truth, who is innocent and who is guilty?

(c) Assuming that the innocent told the truth and the guilty lied, who is innocent and who is guilty?

[Hint: Let H denote: Horace is innocent.
Let G denote: Gladstone is innocent.
Let K denote: Klunker is innocent.
Now symbolize all three testimonies and make one single truth table with a column for each testimony. The desired information can be read from the table.]

5. Agent 006 knows that exactly one of four diplomats is really a spy. He interrogates them, and they make the following statements:

Diplomat A: Diplomat C is the spy.
Diplomat B: I am not a spy.
Diplomat C: Diplomat A's statement is false.
Diplomat D: Diplomat A is the spy.

(a) If 006 knows that exactly one diplomat is telling the truth, who is the spy?

(b) If 006 knows that exactly one diplomat's statement is false, who is the spy?

[Hint: Let A denote: Diplomat A is a spy.
Let B denote: Diplomat B is a spy.
Let C denote: Diplomat C is a spy.
Let D denote: Diplomat D is a spy.
Now symbolize the reply of each diplomat and make a single truth table with a column for each reply. The desired information can be read from the table.]

***6.** A certain college offered exactly four languages: French, German, Russian, and Latin. The registrar was instructed to enroll each student for exactly two languages. After registration, the following facts were compiled:

i) All students who registered for French also registered for exactly one of the other three languages.

ii) All students who registered for neither Latin nor German registered for French.

iii) All students who did not register for Russian registered for *at least* two of the other three languages.

iv) No candidate who registered for Latin and German registered for Russian.

Did the registrar actually follow his instructions?

---

*This denotes a difficult problem.

[Hint: Let x be an arbitrary student at the college.
Let F denote: x registered for French.
Let G denote: x registered for German.
Let L denote: x registered for Latin.
Let R denote: x registered for Russian.
Symbolize each of the four facts. Make a truth table with a column for each of the four facts. Locate the rows for which all four facts have truth value t. Examine these rows closely.]

1.6 Tautologies, Contradictions, and Contingencies

Certain logic forms have truth tables in which the right-hand column consists solely of t's. These forms are called **tautologies.** Hence a tautology is a logic form which has truth value t no matter what values are given to its variable symbols. If the right hand column of the truth table of a logic form consists solely of f's, the logic form is called a **contradiction.** Thus a contradiction has truth value f no matter what values are given to its variable symbols. If a logic form is not a tautology and not a contradiction, it is called a **contingency.**

Example 1. Tautology: $A \vee \neg A$
Contradiction: $A \wedge \neg A$
Contingency: $A \rightarrow B$

Notice that if the variable symbols of a tautology are replaced by sentences, the compound sentence is *always* true.

Example 2. You have already observed that $A \vee \neg A$ is a tautology. Thus the sentence:

$$\underbrace{\text{It is raining}}_{A} \text{ or } \underbrace{\text{it is not raining}}_{\neg A}$$

is a true sentence.

The truth of the sentence in this example does not depend on any facts from meteorology. It is true by virtue of the way it is built up from its parts by means of the logical operations. In other words, it is true by virtue of its form. Logicians call a sentence like this one a **substitutive instance** of a tautology.

Exercises 1.6

1. Make truth tables for the following logic forms. Indicate which are tautologies, which are contradictions, and which are contingencies.

(a) $A \rightarrow A$ (b) $A \rightarrow (B \rightarrow A)$
(c) $(A \vee B) \rightarrow (A \wedge \neg C)$ (d) $A \vee (\neg A \wedge C)$
(e) $A \rightarrow (\neg A \wedge B)$ (f) $\neg (A \wedge B) \leftrightarrow (\neg A \vee \neg B)$

1.7
Sentential Inconsistency

Let $\mathbf{A}_1, \mathbf{A}_2 \ldots \mathbf{A}_n$ be a collection of logic forms. These logic forms are said to be **sententially inconsistent** if their conjunction is a contradiction. Otherwise, they are said to be **sententially consistent.**

To test a collection of logic forms $\mathbf{A}_1, \mathbf{A}_2 \ldots \mathbf{A}_n$ for sentential inconsistency, you have only to form their conjunction $\mathbf{A}_1 \wedge \mathbf{A}_2 \wedge \cdots \wedge \mathbf{A}_n$, make a truth table for this conjunction, and look to see whether the right-hand column of the table has all f's. If this is so, then they are sententially inconsistent. If there is at least one t, then they are sententially consistent.

If, after symbolizing a collection of sentences, you find that the resulting collection of logic forms is sententially inconsistent, then you would know that these sentences are built from their elementary sentences in such a way that it is impossible for all of them to be true.

Exercises 1.7

1. Test the following collection of logic forms for sentential inconsistency:

$\neg (A \wedge B) \quad \neg (A \wedge \neg B) \quad \neg (\neg A \wedge B) \quad \neg (\neg A \wedge \neg B)$

2. Symbolize the following sentences and test the resulting logic forms for sentential inconsistency.

(a) Either a recession will occur or, if unemployment does not decrease, wage controls will be imposed. If a recession does not occur, unemployment will decrease. If wage controls are imposed, unemployment will not decrease.

(b) If imports increase or exports decrease, either tariffs are imposed or devaluation occurs. Tariffs are imposed when and only when imports increase and devaluation does not occur. If exports decrease, then tariffs are not imposed or imports do not increase. Either devaluation does not occur, or tariffs are imposed and exports decrease.

3. After discovering his immense popularity with the American people, Bagel the beagle decided to run for president. He called together his top political advisors for a brainstorming session. Out of this session came the following advice:

Owl: Beagles can't be president, or the country will go to the dogs.
Fox: If a beagle can be president, the country won't go to the dogs.
Bear: Either a beagle can be president, or the country will go the dogs.
Cat: It's not the case that (the country will go to the dogs and a beagle can't be president).

Since Bagel's confidence in his advisors was a shade less than absolute, he decided to run a sentential consistency test. What did it reveal?

1.8
Constructing Logic Forms from Truth Tables

You know that for each logic form a truth table can be made. In this section you will learn how to reverse the procedure. Here we have a truth table for an unknown logic form with n given variable symbols. The problem will be to construct a logic form having this given table for its truth table.

Given the truth table, the following procedure is employed to construct the desired logic form.

1. For each row whose right hand entry is t, make a check mark ✓ to the right of that row.

2. To the right of each check mark √ write a sequence of n terms (one corresponding to each variable symbol) as follows: if in that row a variable symbol has the value t entered for it, then the corresponding term is to be that variable symbol itself; if the variable symbol has the value f entered for it, then the corresponding term is to be the *negation* of that variable symbol.
3. To the right of each sequence of terms which you made in step 2, write the conjunction of those terms.
4. Beneath the table, form the disjunction of all the conjunctions which you made in step 3.

The disjunction formed in step 4 will be the desired answer.

Example. Construct a logic form having the following truth table.

A	B	C	???
t	t	t	f
t	t	f	t
t	f	t	t
t	f	f	f
f	t	t	f
f	t	f	f
f	f	t	t
f	f	f	f

ANSWER:

A	B	C	???	Step 1	Step 2	Step 3
t	t	t	f			
t	t	f	t	√	$A, B, \neg C$	$A \wedge \neg B \wedge C$
t	f	t	t	√	$A, \neg B, C$	$A \wedge B \wedge \neg C$
t	f	f	f			
f	t	t	f			
f	t	f	f			
f	f	t	t	√	$\neg A, \neg B, C$	$\neg A \wedge \neg B \wedge C$
f	f	f	f			

Step 4. $(A \wedge B \wedge \neg C) \vee (A \wedge \neg B \wedge C) \vee (\neg A \wedge \neg B \wedge C)$

The result of Step 4 is the desired logic form. To check your work, make the truth table for the logic form just constructed.

A	B	C	$(A \wedge B \wedge \neg C) \vee (A \wedge \neg B \wedge C) \vee (\neg A \wedge \neg B \wedge C)$
t	t	t	f
t	t	f	t
t	f	t	t
t	f	f	f
f	t	t	f
f	t	f	f
f	f	t	t
f	f	f	f

Comparing this table with the table you were given, you can see that they coincide. Hence

$$(A \wedge B \wedge \neg C) \vee (A \wedge \neg B \wedge C) \vee (\neg A \wedge \neg B \wedge C)$$

is a correct answer.

Probably you have noticed that this method doesn't mention what to do in case the right-hand column of the given table only has f's. This case is even easier; just make a contradiction of the form $(\mathbf{X} \wedge \neg \mathbf{X})$ for each variable symbol $\mathbf{X}$, and then take the disjunction of these contradictions.

Exercises 1.8

1. Using the procedure outlined in this section, find a logic form for each of the following truth tables:

(a)

A	B	??
t	t	t
t	f	t
f	t	t
f	f	f

(b)

A	B	??
t	t	f
t	f	t
f	t	f
f	f	t

(c)

A	B	C	??
t	t	t	f
t	t	f	f
t	f	t	t
t	f	f	t
f	t	t	t
f	t	f	f
f	f	t	f
f	f	f	t

(d)

A	B	C	??
t	t	t	t
t	t	f	t
t	f	t	t
t	f	f	f
f	t	t	t
f	t	f	t
f	f	t	t
f	f	f	f

Check each answer by making the truth table of each logic form constructed.

2. Find a logic form with the 3 variable symbols A, B, C, which has the following truth table:

A	B	C	???
t	t	t	f
t	t	f	f
t	f	t	f
t	f	f	f
f	t	t	f
f	t	f	f
f	f	t	f
f	f	f	f

***3.** The people who live on the banks of the Ooga River always tell the truth or always tell lies, and they respond to questions only with a yes or a no. An explorer comes to a fork in the river, where one branch leads to the top of a mile high waterfall and the other branch leads to a settlement. Of course there is no sign telling which branch leads to the settlement, but there is a native, Mr. Blanco, standing on the

*This denotes a difficult problem.

shore. What yes-or-no question should the explorer ask Mr. Blanco to determine which branch leads to the settlement?

[Hint: Let A stand for "Mr. Blanco tells the truth," and let B stand for "The left-hand branch leads to the settlement." Construct, by means of a suitable truth table, a logic form involving A and B such that Blanco's answer to the question will be "yes" (i.e., true) if and only if B is true.]

2

Algebra of Logic

2.1
Logical Equivalence

Let **A** and **B** be logic forms. **A** and **B** are **logically equivalent** if and only if the logic form $\mathbf{A} \leftrightarrow \mathbf{B}$ is a tautology. If **A** is logically equivalent to **B**, it is indicated by writing $\mathbf{A} \equiv \mathbf{B}$.

Suppose $\mathbf{A} \equiv \mathbf{B}$. Then if all the variable symbols appearing in **A** or **B** are listed, the assignments of truth values to these symbols which make **A** true will be precisely those which make **B** true.

Example 1. Prove: $(A \rightarrow B) \equiv \neg (A \wedge \neg B)$

ANSWER: To do this, construct the truth table of the logic form

$$(A \rightarrow B) \leftrightarrow \neg (A \wedge \neg B)$$

which, for brevity, will be denoted by **E**.

A	B	A → B	¬(A ∧ ¬B)	E
t	t	t	t	t
t	f	f	f	t
f	t	t	t	t
f	f	t	t	t

From the truth table you can see that **E** is a tautology. Hence:

$$A \rightarrow B \equiv \neg (A \wedge \neg B)$$

Example 2. Prove: $A \rightarrow (B \rightarrow A) \equiv C \vee \neg C$

ANSWER: Construct the truth table for the logic form

$$[A \rightarrow (B \rightarrow A)] \leftrightarrow (C \vee \neg C)$$

which is denoted by **E**.

A	B	C	$A \rightarrow (B \rightarrow A)$	$C \vee \neg C$	E
t	t	t	t	t	t
t	t	f	t	t	t
t	f	t	t	t	t
t	f	f	t	t	t
f	t	t	t	t	t
f	t	f	t	t	t
f	f	t	t	t	t
f	f	f	t	t	t

By inspecting the truth table for **E**, you can conclude:

$$A \rightarrow (B \rightarrow A) \equiv (C \vee \neg C)$$

The second example shows that the tautologies $A \rightarrow (B \rightarrow A)$ and $C \vee \neg C$ are logically equivalent. Actually it is true that all tautologies are logically equivalent, and all contradictions are logically equivalent.

Exercises 2.1

1. By constructing the appropriate truth table, prove or disprove the following claims:

(a) $(A \vee B) \equiv \neg(\neg A \wedge \neg B)$
(b) $(A \vee B) \equiv \neg(\neg B \wedge \neg C)$
(c) $A \wedge (B \vee C) \equiv (A \wedge B) \vee (A \wedge C)$
(d) $C \wedge (B \vee D) \equiv (C \vee D) \wedge (B \vee D)$
(e) $A \rightarrow (B \rightarrow C) \equiv (A \rightarrow B) \rightarrow C$

2. Construct a logic form distinct from, but logically equivalent to, the logic form

$$A \rightarrow (B \rightarrow C).$$

[Hint: Make the truth table for $A \rightarrow (B \rightarrow C)$. Then use the procedure for constructing a logic form, given its truth table. Check that the logic form so obtained is logically equivalent to $A \rightarrow (B \rightarrow C)$.]

2.2 Basic Equivalences

Here is a list of twenty basic equivalences which will be used repeatedly throughout this book. You should familiarize yourself with all of the equivalences listed. They can be easily verified by the truth table technique which you used earlier to check for logical equivalency. (In fact, you already verified a couple of them in previous exercises.)

The symbol **T** is used to denote the particular tautology $A \vee \neg A$, and the symbol **F** is used to denote the particular contradiction $A \wedge \neg A$.

Basic Equivalences

Equivalence	Name
1. $\neg \neg A \equiv A$	law of double negation
2. $A \vee B \equiv B \vee A$ 3. $A \wedge B \equiv B \wedge A$	commutative laws
4. $A \vee (B \vee C) \equiv (A \vee B) \vee C$ 5. $A \wedge (B \wedge C) \equiv (A \wedge B) \wedge C$	associative laws
6. $A \vee (B \wedge C) \equiv (A \vee B) \wedge (A \vee C)$ 7. $A \wedge (B \vee C) \equiv (A \wedge B) \vee (A \wedge C)$	distributive laws
8. $\neg (A \vee B) \equiv \neg A \wedge \neg B$ 9. $\neg (A \wedge B) \equiv \neg A \vee \neg B$	de Morgan's laws
10. $A \wedge A \equiv A$ 11. $A \vee A \equiv A$	idempotent laws
12. $A \wedge (A \vee B) \equiv A$ 13. $A \vee (A \wedge B) \equiv A$	absorption laws
14. $B \wedge T \equiv B$ 15. $B \vee F \equiv B$	identity laws
16. $B \vee T \equiv T$ 17. $B \wedge F \equiv F$	domination laws
18. $A \rightarrow B \equiv \neg A \vee B$	arrow law
19. $A \rightarrow B \equiv \neg B \rightarrow \neg A$	law of contraposition
20. $A \leftrightarrow B \equiv (A \rightarrow B) \wedge (B \rightarrow A)$	double arrow law

We close this section with a few observations. In Chapter 1 we introduced the concept of "contrapositive" of an implication. Basic equivalence 19 asserts that an implication is logically equivalent to its contrapositive.

From your basic algebra course you may be familiar with the commutative, associative, and distributive laws. In contrast to the situation in ordinary algebra, there are two distributive laws here. Note that one of these may be obtained from the other by interchanging the $\wedge$ and $\vee$ symbols. In ordinary algebra this doesn't work. If you interchange the + and · symbols in the distributive law

$$a \cdot (b + c) = (a \cdot b) + (a \cdot c)$$

you get

$$a + (b \cdot c) = (a + b) \cdot (a + c)$$

which is false in ordinary algebra. (Note, for example, that it fails when a, b, c, each have the value 1.)

2.3
Algebraic Manipulation

In this section you will learn another method for proving that two logic forms are logically equivalent. For this procedure you need the following two rules.

I. **Replacement rule.** If part of a logic form **A** is replaced at one or more occurrences by a logically equivalent form, then the result is logically equivalent to the original form **A**.

II. Transitivity rule. If **A, B,** and **C** are logic forms such that $\mathbf{A} \equiv \mathbf{B}$ and $\mathbf{B} \equiv \mathbf{C}$, then $\mathbf{A} \equiv \mathbf{C}$.

The technique used to show logical equivalency here resembles the process used in algebra to show that two algebraic expressions are equal. Starting with one of the logic forms and making successive applications of the replacement rule and the basic equivalences of the preceding section you convert it into the other logic form. Then, by using the transitivity rule, you can conclude that the two logic forms are logically equivalent.

Here are a few examples to show the algebraic manipulation of logic forms. Study these examples carefully before attempting the exercises which follow.

Example 1. Prove: $A \wedge (B \vee \neg A) \equiv A \wedge B$

PROOF: Start with the left side and, in a series of steps (listed vertically), manipulate your way to the right side:

$A \wedge (B \vee \neg A)$

distributive law

$(A \wedge B) \vee (A \wedge \neg A)$

definition of **F**

$(A \wedge B) \vee \mathbf{F}$

identity law

$A \wedge B$

Since you went from the left side to the right side in steps which leave each line logically equivalent to the one preceding it, you can conclude by the transitivity rule that the desired equivalence holds.

Example 2. Show that $\neg(A \rightarrow B) \equiv A \wedge \neg B$

ANSWER: $\neg (A \rightarrow B)$

arrow law

$\neg (\neg A \vee B)$

de Morgan's law

$\neg \neg A \wedge \neg B$

double negation law

$A \wedge \neg B$

Since you started with the left side of the equivalence in question and ended with the right side, you can conclude by the transitivity rule that the equivalence holds.

Example 3. Simplify $[(B \vee \neg A) \wedge B] \rightarrow B$

ANSWER: $[(B \vee \neg A) \wedge B] \rightarrow B$

commutative law

$[B \wedge (B \vee \neg A)] \rightarrow B$

absorption law

$B \rightarrow B$

Example 4. Write the negation of the following sentence, and then express it in such a way that the negated parts are elementary (not compound) sentences

If the survey is accurate, we should propose legislation to Congress and release our findings to the newspapers.

ANSWER: Let I be: The survey is accurate.
Let L be: We should propose legislation to Congress.
Let N be: We should release our findings to the newspapers.
Then you can represent the negation of the given sentence symbolically as:

$\neg [I \rightarrow (L \wedge N)]$

You can now manipulate this logic form:

$\neg [I \rightarrow (L \wedge N)]$

arrow law

$\neg [\neg I \vee (L \wedge N)]$

de Morgan's law

$\neg\neg I \wedge \neg (L \wedge N)$

double negation law

$I \wedge \neg (L \wedge N)$

de Morgan's law

$I \wedge (\neg L \vee \neg N)$

Hence the negation of the given sentence, can be written: The survey is accurate and either we should not propose legislation to Congress or we should not release our findings to the newspapers.

Exercises 2.3

1. Establish the following equivalences using algebraic manipulations.

(a) $A \vee B \equiv \neg A \rightarrow B$
(b) $\neg (A \rightarrow B) \equiv A \wedge \neg B$
(c) $A \rightarrow (B \rightarrow C) \equiv (A \wedge B) \rightarrow C$
(d) $(A \wedge B) \vee (\neg A \wedge C) \equiv (A \rightarrow B) \wedge [C \vee (A \wedge B)]$

2. Simplify the following expressions. (Try to minimize the total number of occurrences of operation symbols and variable symbols. Answers won't be unique.)

(a) $\neg A \vee (A \vee B)$
(b) $\neg (\neg B \vee A)$
(c) $\neg A \wedge (A \vee B)$
(d) $(A \wedge B) \vee (A \wedge \neg B)$
(e) $\neg (A \wedge \neg (A \vee B))$
(f) $A \wedge (A \vee B \vee C)$
(g) $[(\neg B \vee A) \wedge A] \rightarrow A$
(h) $(A \vee B) \wedge [(B \vee A) \wedge C]$
(i) $\neg (A \vee \neg (B \wedge C))$
(j) $A \wedge (\neg A \vee B)$
(k) $A \vee [\neg A \vee (B \wedge \neg C)]$
(l) $[(\neg A \wedge \neg B) \vee \neg (A \vee B)] \rightarrow (\neg A \wedge \neg B)$
(m) $[A \wedge (B \wedge C)] \vee [\neg A \wedge (B \wedge C)]$
(n) $E \wedge D \wedge [\neg [(A \rightarrow B) \leftrightarrow (\neg B \rightarrow \neg A)] \rightarrow C]$

3. Negate each of the following sentences, and then express the answer in such a way that only elementary (not compound) parts of the sentence are negated.

(a) Freedom of the press is an important safeguard of liberty, and in protecting it, our courts have played a major role.

(b) If a politician seeks the presidency, and he has sufficient financial backing, then he can afford to appear on nationwide television frequently.

(c) A man has self respect if he is contributing to a better society.

(d) We can halt pollution only if we act now.

(e) If a man cannot join the union, he cannot find a job in that factory and he must relocate his family.

(f) If a worker can share on the company profits, he works harder and he demands fewer fringe benefits.

4. If $\mathbf{A} \wedge \mathbf{B} \equiv \mathbf{A} \wedge \mathbf{C}$, can you conclude that $\mathbf{B} \equiv \mathbf{C}$?

5. Show that you can manipulate from one side of the absorption law to the other side, using only the remaining 18 basic equivalences.

6. The Internal Revenue Service lists the following three rules as guidelines for filing tax returns:

(a) A person pays taxes only if he is over 18 years of age, or has earned $1000 during the past twelve months, or both.

(b) No widow must pay taxes unless she has earned $1000 during the past twelve months.

(c) Anyone over 18 years of age who has not earned $1000 during the past twelve months does not pay any taxes.

Find a simpler form for these rules.

[Hint: Let x be an arbitrary person.
Let T denote: x is a taxpayer.
Let W denote: x is a widow.
Let A denote: x is over 18 years of age.
Let E denote: x earned over $1000 in the past twelve months.
Then the guidelines given can be represented symbolically by:

$$[T \rightarrow (A \vee E)] \wedge [\neg E \rightarrow \neg (W \wedge T)] \wedge [(A \wedge \neg E) \rightarrow \neg T]$$

Simplify this logic form.]

***7.** In the Tooba tribe, the elders use the following rules to decide a man's role in the tribal society. Try to replace this set of rules by a simpler (but logically equivalent) set of rules.

(a) The warriors shall be chosen from among the hunters.

(b) Any Tooba man who is both a farmer and a hunter should also be a warrior.

(c) No farmer shall be a warrior.

[Hint: Let x be an arbitrary Tooba man.
Let W denote: x is a warrior.
Let F denote: x is a farmer
Let H denote: x is a hunter.]

---

*This denotes a difficult problem.

2.4 Conjunctive Normal Form

Logic form **A** is a **conjunctive normal form** if it satisfies one of the following conditions:

i) **A** is a variable symbol.
ii) **A** is the negation of a variable symbol.
iii) **A** is a disjunction of two or more terms, each of which is either a variable symbol or the negation of a variable symbol.
iv) **A** is a conjunction of two or more terms, each of which is one of the three types listed above.

It is very important to be able to recognize a conjunctive normal form.

Example 1. The following logic forms are conjunctive forms:

A (satisfies condition i)

$\neg B$ (satisfies condition ii)

$\neg C \vee B \vee A$ (satisfies condition iii)

$(\neg C \vee B \vee A) \wedge A \wedge \neg B$ (satisfies condition iv)

$(\neg A \vee B) \wedge (C \vee A) \wedge (B \vee C)$ (satisfies condition iv)

The following logic forms are *not* conjunctive normal forms:

$A \rightarrow B$

$(A \wedge B) \vee \neg C$

$(A \vee B) \wedge (\neg B \rightarrow A)$

$\neg (A \wedge B)$

To help fix the idea in your mind, think of a conjunctive normal form as a logic form which looks like

$$(\mathbf{A} \vee \mathbf{B} \vee \cdots) \wedge (\mathbf{C} \vee \mathbf{D} \vee \cdots) \wedge \cdots \wedge (\mathbf{E} \vee \mathbf{F} \vee \cdots)$$

where the heavy-type letters denote logic forms no more complicated than single variable symbols or negations of variable symbols. We are allowing as special cases of this, the case where only one of the bracketed expressions actually appears, and the case where some of the bracketed expressions have only one term. The bracketed expressions between the conjunction symbols are called the **conjuncts** of the normal form. If the conjunctive normal form has no conjunction symbols, the entire expression is thought of as a single conjunct. Thus, $A \vee B \vee \neg C$ has only one conjunct, namely the expression $A \vee B \vee \neg C$.

Exercises 2.4

1. Determine which (if any) of the following logic forms are conjunctive normal forms.

(a) $(A \vee B \vee \neg B) \wedge (A \vee C)$
(b) $(A \vee B \vee \neg B) \wedge (A \wedge C)$
(c) $(A \rightarrow B) \vee (C \wedge \neg B)$
(d) $A \wedge B$
(e) $\neg A \wedge B$
(f) $\neg [(A \vee B) \wedge (A \vee C)]$
(g) $\neg A \wedge B \wedge \neg C$
(h) $A \wedge [B \vee (C \wedge \neg D)]$
(i) $(A \wedge B) \vee [(A \wedge C) \vee (A \wedge \neg B)]$

2.5
Reduction to Conjunctive Normal Form

By using algebraic manipulations it is possible to reduce any logic form to an equivalent conjunctive normal form.

Example. Find a conjunctive normal form logically equivalent to:

$(A \wedge B) \vee (B \wedge \neg C)$

ANSWER: $(A \wedge B) \vee (B \wedge \neg C)$

distributive law

$[(A \wedge B) \vee B] \wedge [(A \wedge B) \vee \neg C]$

distributive law

$[(A \vee B) \wedge (B \vee B)] \wedge [(A \wedge B) \vee \neg C]$

distributive law

$[(A \vee B) \wedge (B \vee B)] \wedge [(A \vee \neg C) \wedge (B \vee \neg C)]$

associative law

$(A \vee B) \wedge (B \vee B) \wedge (A \wedge \neg C) \wedge (B \vee \neg C)$

The last line is, as desired, a conjunctive normal form. It is not the simplest answer, but for the intended applications of conjunctive normal form, this will not matter.

Note that a logic form might be manipulated into a conjunctive normal form in various ways, and the answer might come out in various ways. All answers, of course, would be logically equivalent.

Now it turns out that in reducing to a conjunctive normal form you need to use just one of the two distributive laws learned in Section 2.2. If you now switch to an "arithmetical notation" and write

$\mathbf{A} + \mathbf{B}$ instead of $\mathbf{A} \wedge \mathbf{B}$

$\mathbf{AB}$ instead of $\mathbf{A} \vee \mathbf{B}$

then you see that of the two distributive laws, the one you need appears in the new notation as

$$\mathbf{A}(\mathbf{B} + \mathbf{C}) \equiv \mathbf{AB} + \mathbf{AC}.$$

The other distributive law would look rather strange and unfamiliar in arithmetical notation. It would appear as

$$\mathbf{A} + (\mathbf{BC}) \equiv (\mathbf{A} + \mathbf{B})(\mathbf{A} + \mathbf{C}).$$

You are probably so accustomed to working with ordinary numbers where this second distributive law fails, that we might have confused you if we had introduced and used arithematical notation earlier in the book. Fortunately, for reduction to conjunctive normal form you do not need to make use of this "funny looking" distributive law. So perhaps in reducing to conjunctive normal form, it might be more convenient to use arithmetical notation. For one thing, your expressions will take up less space. For another, repeated application of the distributive law to complicated expressions can be handled just as though you were "multiplying out" an expression of ordinary algebra.

A further space saving device would be to write

$\mathbf{A}'$ instead of $\neg \mathbf{A}$.

De Morgan's laws would then appear as

$$(\mathbf{A} + \mathbf{B})' \equiv \mathbf{A}'\mathbf{B}'$$

$$(\mathbf{AB})' \equiv \mathbf{A}' + \mathbf{B}'$$

Of course, for uncomplicated expressions you may prefer not to use the new arithmetical notation at all. It is entirely optional.

A good way to reduce to a conjunctive normal form is to perform the following four steps:

1. Use the arrow laws to remove all $\rightarrow$ and $\leftrightarrow$.
2. Use de Morgan's laws repeatedly until the only negated terms are variable symbols.
3. Switch to arithmetical notation.
4. Multiply out (as in ordinary algebra).

Naturally, if you get a chance to simplify things along the way by using the absorption, idempotent, domination, or identity laws, so much the better.

Example. Find a conjunctive normal form logically equivalent to:

$$[(A \rightarrow B) \rightarrow C] \vee [D \wedge E]$$

ANSWER:

$$[(A \rightarrow B) \rightarrow C] \vee [D \wedge E]$$

arrow law

$$[(\neg A \vee B) \rightarrow C] \vee [D \wedge E]$$

arrow law

$$[\neg (\neg A \vee B) \vee C] \vee [D \wedge E]$$

de Morgan's law

$$[(\neg \neg A \wedge \neg B) \vee C] \vee [D \wedge E]$$

double negation law

$$[(A \wedge \neg B) \vee C] \vee [D \wedge E]$$

change notation to arithmetical

$$[(A + B')C](D + E)$$

associative and commutative laws

$$C(A + B')(D + E)$$

multiply out (distributive law)

$$CAD + CB'D + CAE + CB'E$$

The last line is the desired conjunctive normal form.

A good way to remember that it is the "and" symbol $\wedge$ which corresponds to the symbol + when reducing to a conjunctive normal form, is to think of the word CAP, which is made from the first letters of the words

Conjunctive (normal form) And Plus.

Exercises 2.5

1. Reduce each of the following logic forms to a conjunctive normal form:

(a) $\neg A \rightarrow B$
(b) $\neg(\neg B \vee A)$
(c) $(A \wedge B) \vee (A \wedge \neg B)$
(d) $\neg(A \wedge \neg(A \vee B))$
(e) $A \leftrightarrow B$
(f) $A \wedge [B \vee (C \wedge \neg D)]$
(g) $(A \rightarrow B) \wedge [C \vee (A \wedge B)]$
(h) $(A \wedge B) \vee [(A \wedge C) \vee (A \wedge \neg B)]$

2.6 Uses of Conjunctive Normal Form

If a logic form **A** is a conjunctive normal form, there is a way of telling at a glance whether or not it is a tautology. Just apply the following rule:

> A conjunctive normal form is a tautology if and only if in each conjunct there appears some variable symbol and its negation.

In applying the above rule, do not forget that if no conjunction symbols actually appear in the conjunctive normal form, the entire normal form is thought of as a single conjunct.

Example 1. $(A \vee B \vee \neg A) \wedge (C \vee D \vee E \vee \neg D)$ is a tautology because of the appearance of A, $\neg A$ in the left conjunct and D, $\neg D$ in the right conjunct.

$(A \vee B \vee \neg A) \wedge (C \vee \neg B \vee D) \wedge (E \vee \neg E)$ is not a tautology because the middle conjunct fails to have some variable symbol and its negation.

$A \vee B \vee D \vee \neg B$ is a tautology because of the appearance of B, $\neg B$.

$A \vee B \vee \neg C$ is not a tautology. Its only conjunct (the whole expression itself) fails to have some variable symbol and its negation.

Here is how to test *any* logic form **A** for tautology (regardless of whether **A** is a conjunctive normal form or not). First reduce **A** to a conjunctive normal form. Then you can see at a glance whether the normal form (and hence **A**) is a tautology.

Example 2. Is the following logic form a tautology?

$$A \to (B \to A)$$

ANSWER: Reduce to a conjunctive normal form

$$\begin{aligned} A \to (B \to A) &\equiv \neg A \vee (B \to A) \\ &\equiv \neg A \vee (\neg B \vee A) \\ &\equiv \neg A \vee \neg B \vee A \end{aligned}$$

Now $\neg A \vee \neg B \vee A$ is a conjunctive normal form, and it is evident by inspection that it is a tautology. Consequently, so is $A \to (B \to A)$.

Example 3. Is the following logic form a tautology?

$$(A \to B) \to A$$

ANSWER: Reduce to a conjunctive normal form

$$\begin{aligned} (A \to B) \to A &\equiv \neg(\neg A \vee B) \vee A \\ &\equiv (\neg\neg A \wedge \neg B) \vee A \\ &\equiv (\neg\neg A \vee A) \wedge (\neg B \vee A) \\ &\equiv (A \vee A) \wedge (\neg B \vee A) \\ &\equiv A \wedge (\neg B \vee A) \end{aligned}$$

Now $A \wedge (\neg B \vee A)$ is a conjunctive normal form. By inspection it is evident that it is not a tautology, so neither is $(A \to B) \to A$.

In both examples a truth table could have been used just as easily to settle the question. For expressions with three or more distinct letters, making a table can be tedious, and the method of conjunctive normal forms is preferable.

Using a conjunctive normal form to test for tautologies also gives additional information. If the test shows that a logic form is not a tautology, the conjunctive normal form helps to find an assignment of truth values to the variable symbols, which makes the original logic form have truth value f. All that has to be done is to locate a conjunct which fails to have both a variable symbol and its negation. Then assign values to the variables so as to make that conjunct false.

Exercises 2.6

1. Test the following to see if they are tautologies using the method of conjunctive normal form. For each logic form which is not a tautology, use the conjunctive normal form to find an assignment of truth values to the variables which makes the logic form false.

(a) $[A \wedge (A \rightarrow B)] \rightarrow B$
(b) $A \vee \neg A$
(c) $A \rightarrow (A \vee B)$
(d) $[(A \rightarrow B) \wedge \neg A] \rightarrow B$
(e) $[\neg (A \rightarrow B)] \rightarrow (A \vee B)$
(f) $[A \vee B] \rightarrow A$
(g) $(A \rightarrow B) \vee (C \wedge B)$
(h) $(A \wedge B) \vee [(D \rightarrow C) \vee E]$
(i) $(A \leftrightarrow B) \wedge [(C \vee D) \rightarrow A]$

2.7 Disjunctive Normal Form

A logic form **A** is a **disjunctive normal form** if its satisfies one of the following conditions:

i) **A** is a variable symbol.
ii) **A** is the negation of a variable symbol.
iii) **A** is a conjunction of two or more terms each of which is a variable symbol or the negation of a variable symbol.
iv) **A** is a disjunction of two or more terms each of which is one of the three types described above.

Example 1. The following logic forms are disjunctive normal forms:

A

$\neg B \vee C$

$A \wedge B \wedge \neg C \wedge D$

$(\neg B \wedge C) \vee (A \wedge D) \vee \neg E$

The following logic forms are not disjunctive normal forms:

$A \rightarrow B$

$(\neg B \vee C) \wedge A$

$(A \wedge C) \vee (A \rightarrow B)$

Note that some logic forms, for example the form $\neg A$, can be both a conjunctive normal form *and* a disjunctive normal form.

A disjunctive normal form looks just like a conjunctive normal form except that the roles of the symbols $\wedge$ and $\vee$ are reversed. Thus

$$(A \vee \neg B) \wedge D$$

is a conjunctive form, while

$$(A \wedge \neg B) \vee D$$

is a disjunctive normal form.

Every logic form **A** can be reduced to a logic form which is a disjunctive normal form logically equivalent to **A**.

Example 2. Find a disjunctive normal form which is logically equivalent to:

$$(A \rightarrow B) \rightarrow C$$

ANSWER: $(A \rightarrow B) \rightarrow C$

arrow law

$\neg (A \rightarrow B) \vee C$

arrow law

$\neg (\neg A \vee B) \vee C$

de Morgan's law

$(A \wedge \neg B) \vee C$

The last line is, as desired, a disjunctive normal form.

Example 3. Find a disjunctive normal form which is logically equivalent to:

$A \wedge (B \vee C)$

ANSWER: $A \wedge (B \vee C) \equiv (A \wedge B) \vee (A \wedge C)$

The right-hand side is the answer.

Now it turns out that in reducing to disjunctive normal form you need to use only one of the two distributive laws. In fact, it's the one you did not need for reducing to conjunctive normal form. Recall that in switching to arithmetical notation this distributive law looked unfamiliar. You can avoid dealing with a "funny looking" representation of this distributive law by reversing the role of the + symbol. This time write

$\mathbf{A + B}$ for $\mathbf{A \vee B}$

$\mathbf{AB}$ for $\mathbf{A \wedge B}$

Again the distributive law that you do need is

$$\mathbf{A(B + C) \equiv AB + AC}.$$

Any logic form can be reduced to a disjunctive normal form by the following steps:

1. Eliminate arrows.
2. Repeat the use of de Morgan's laws until negation symbols appear only in front of variable symbols.
3. Switch to arithmetical notation.
4. Multiply out (like in ordinary algebra).

In order to help you remember that it is the "or" symbol which should correspond to + in reducing to a disjunctive normal form, think of the word DOPE whose first three letters come from

Disjunctive (normal form) **O**r **P**lus.

Recall that for conjunctive normal form the key word was CAP. Of course, if the logic form in question isn't very complicated, you may prefer not to use arithmetical notation.

Example 4. Find a disjunctive normal form logically equivalent to

$\neg(A \wedge B) \wedge (A \rightarrow C) \wedge C$

ANSWER: $\neg(A \wedge B) \wedge (A \rightarrow\ ) \wedge C$

arrow law

$\neg(A \wedge B) \wedge (\neg A \vee C) \wedge C$

de Morgan's law

$(\neg A \vee \neg B) \wedge (\neg A \vee C) \wedge C$

change notation to arithmetical

$(A' + B')(A' + C)C$

multiply out (distributive law)

$A'A'C + B'A'C + A'CC + B'CC.$

The last line is a disjunctive normal form.

Exercises 2.7

1. Determine which (if any) of the following logic forms are disjunctive normal forms.

(a) $A \wedge B$ (b) $A \vee B$
(c) $A \wedge (B \vee C)$ (d) $A \vee (B \wedge C)$
(e) $A \vee \neg C \vee (A \wedge C)$ (f) $(A \rightarrow B) \vee \neg C$
(g) $\neg (A \vee B)$ (h) $\neg A \wedge \neg B$

2. Reduce each of the following forms to a logic form which is a disjunctive normal form.

(a) $A \rightarrow B$ (b) $A \vee B$
(c) $\neg (A \vee B)$ (d) $(A \leftrightarrow B)$
(e) $A \wedge B \wedge C$ (f) $(A \vee B) \wedge (B \rightarrow C)$
(g) $[\neg (B \rightarrow C)] \rightarrow A$ (h) $\neg [(B \rightarrow C) \rightarrow A]$
(i) $(A \vee B) \wedge [(B \vee A) \wedge C]$
(j) $A \wedge [(\neg A \rightarrow B) \vee C]$
(k) $(A \rightarrow B) \rightarrow [(B \rightarrow C) \rightarrow (A \rightarrow C)]$

2.8
Uses of Disjunctive Normal Form

If a logic form is a disjunctive normal form, there is a way of telling by inspection whether or not it is a contradiction. Just apply the following rule:

A disjunctive normal form is a contradiction if and only if in each disjunct there appears some variable symbol and its negation.

Example 1. $D \wedge \neg D$ is a contradiction.

$(A \wedge B) \vee (A \wedge C)$ is not a contradiction.

B is not a contradiction.

$(A \wedge \neg A) \vee (A \wedge C)$ is not a contradiction.

$(A \wedge \neg A \wedge B)$ is a contradiction.

$(A \wedge \neg A) \vee (B \wedge C \wedge \neg B)$ is a contradiction.

To test an arbitrary logic form to see if it is a contradiction, reduce it to a disjunctive normal form, and then check the normal form by inspection.

Example 2. Use reduction to a disjunctive normal form to tell whether the following logic form is a contradiction.

$(A \rightarrow \neg B) \wedge (A \wedge B)$

ANSWER:

$$(A \rightarrow \neg B) \wedge (A \wedge B) \equiv (\neg A \vee \neg B) \wedge (A \wedge B)$$
$$\equiv (\neg A \wedge A \wedge B) \vee (\neg B \wedge A \wedge B)$$

Each disjunct has some letter and its negation. Therefore, the original logic form is a contradiction.

Even if reduction to a disjunctive normal form reveals that a logic form **A** is not a contradiction, some useful knowledge can be obtained. You can use the disjunctive normal form to determine an assignment of truth values to the variable symbols in **A** which gives **A** truth value t. All you have to do is locate a disjunct which fails to have a variable symbol and its negation. Then assign values to the variables so as to make that disjunct true.

Exercises 2.8

1. Reduce each of the following logic forms to a disjunctive normal form and determine whether it is a contradiction. If it is not a contradiction, use the disjunctive normal form to find an assignment of truth values to the variable symbols which makes the original logic form true.

(a) $A \rightarrow (\neg A \wedge B)$
(b) $A \rightarrow (\neg A \vee B)$
(c) $(A \vee B) \wedge \neg (\neg A \rightarrow B)$
(d) $(A \wedge (B \vee C)) \wedge (\neg A \rightarrow (\neg B \wedge \neg C))$

2. Use reduction to a conjunctive normal form and to a disjunctive normal form to determine which of the following logic forms are tautologies, which are contradictions, and which are contingencies.

(a) $\neg A \vee (B \rightarrow \neg C) \vee (C \rightarrow A)$
(b) $[(A \rightarrow B) \rightarrow C] \rightarrow [(C \rightarrow A) \rightarrow (D \rightarrow A)]$
(c) $(A \rightarrow B) \rightarrow [(B \rightarrow C) \rightarrow (A \rightarrow C)]$
(d) $[(A \wedge C) \vee (B \wedge \neg C)] \leftrightarrow [(\neg A \wedge C) \vee (\neg B \wedge \neg C)]$

3. Check the following set of assumptions for sentential consistency.

If the XYZ company builds a new factory or some of its machinery must be replaced, either the company can obtain a loan or it will be threatened with bankruptcy. The company can obtain a loan if and only if it builds a new factory and no machinery must be replaced. If machinery must be replaced, then the company cannot obtain a loan or it does not build a new factory. Either the company will be threatened with bankruptcy, or some of its machiner must be replaced and the company can obtain a loan.
[Hint: Symbolize the sentences, take the conjunction and reduce to a disjunctive normal form. See if it is a contradiction.]

2.9 Interdependence of the Basic Logical Operations

Previously you were introduced to the five basic logical operations:

$\wedge$
$\vee$
$\neg$
$\rightarrow$
$\leftrightarrow$

The double arrow law mentioned earlier in this chapter shows how to express the operation $\leftrightarrow$ in terms of the operations $\rightarrow$ and $\wedge$:

$$A \leftrightarrow B \equiv (A \rightarrow B) \wedge (B \rightarrow A)$$

It follows that any logic form with occurrences of $\leftrightarrow$ can be converted to a logic form having no occurrences of $\leftrightarrow$ by replacing the left side of the above equivalence with the right side. This shows that you can get along with just the remaining four logical operations. Furthermore, the arrow law and de Morgan's law can be used to express $\rightarrow$ and $\wedge$ in terms of $\vee$ and $\neg$:

$$A \rightarrow B \equiv \neg \ A \vee B$$

$$A \wedge B \equiv \neg [\neg (A \wedge B)] \equiv \neg (\neg A \vee \neg B)$$

Hence, given any logic form A, you can eliminate all occurrences of $\leftrightarrow$, $\rightarrow$, and $\wedge$ in favor of $\vee$ and $\neg$, and the new logic form would be logically equivalency to **A**.

The above discussion shows that of the five basic operations available, it is possible to get along with just two of them, namely $\vee$ and $\neg$.

Example. Find a logic form logically equivalent to the one below, but which uses only the connectives $\vee$ and $\neg$:

ANSWER:

$$\begin{aligned}
A \leftrightarrow B &\equiv (A \rightarrow B) \wedge (B \rightarrow A) \\
&\equiv (\neg A \vee B) \wedge (\neg B \vee A) \\
&\equiv [(\neg A \vee B) \wedge \neg B] \vee [(\neg A \vee B) \wedge A] \\
&\equiv (\neg A \wedge \neg B) \vee (B \wedge \neg B) \\
&\quad \vee [(\neg A \wedge A) \vee (B \wedge A)] \\
&\equiv (\neg A \wedge \neg B) \vee (B \wedge A) \\
&\equiv \neg (A \vee B) \vee \neg (\neg A \vee \neg B).
\end{aligned}$$

The example above shows one reason why we did not merely introduce the two logical operations $\vee$ and $\neg$ at the beginning of the book and do all subsequent work using just these two operations. Many short logic forms, for example $A \leftrightarrow B$, would be expressed by long unwieldy ones. Thus, even though it is possible to manage in principle with just two operations, it is convenient to have all five of them available.

There is a new logical operation in terms of which each of the five basic operations can be expressed. It is called the **Sheffer stroke,** and it is denoted by a vertical stoke $|$. Its truth table is:

TRUTH TABLE FOR $|$
(SHEFFER STROKE)

A	B	A $|$ B
t	t	f
t	f	t
f	t	t
f	f	t

Since the truth table of $A \mid B$ is the same as the truth table of $\neg (A \wedge B)$, we have

$$A \mid B \equiv \neg (A \wedge B).$$

You have already seen that any logic form can be expressed using only the operations $\vee$ and $\neg$. Thus, to conclude that all the basic logical operations can be expressed in terms of the Sheffer stroke alone, you need only show that $\vee$ and $\neg$ can each be expressed in terms of $|$. You can see that $\neg A$ may be so expressed by observing that:

$$\neg A \equiv \neg (A \wedge A) \equiv A \mid A$$

That $A \vee B$ can be expressed in terms of the Sheffer stroke follows from :

$$\begin{aligned} A \vee B &\equiv \neg [\neg (A \vee B)] \\ &\equiv \neg [\neg A \wedge \neg B] \\ &\equiv \neg [(A \mid A) \wedge (B \mid B)] \\ &\equiv (A \mid A) \mid (B \mid B) \end{aligned}$$

The conclusion to be drawn from this section is that every logic form can be expressed by an equivalent one whose sole operation symbol is the Sheffer stroke. Our interest in this result is only theoretical, not practical.

Exercises 2.9

1. For each logic form below, find one logically equivalent to it, but which uses only the operations $\vee$, $\neg$.

(a) $[(A \rightarrow B) \rightarrow C] \wedge A$
(b) $(A \wedge B) \vee (B \rightarrow \neg C)$
(c) $[A \vee (C \wedge D)] \rightarrow B$

2. Show that $\vee$, $\rightarrow$, and $\leftrightarrow$ can each be expressed in terms of $\wedge$ and $\neg$.

3. Find a logic form equivalent to the one below, but which uses only $\wedge$ and $\neg$.

$(A \rightarrow B) \vee C$

4. Express each of the basic logical operations $\vee$, $\wedge$, and $\leftrightarrow$ in terms of $\neg$ and $\rightarrow$.

5. Find a logic form equivalent to the one below but which uses only $\neg$ and $\rightarrow$.

$$A \wedge (B \vee C)$$

6. For each logic form below, find one logically equivalent to it, but which uses only the Sheffer stroke.

(a) $A \wedge B$
(b) $A \rightarrow B$
(c) $A \vee \neg B$
(d) $\neg B \rightarrow A$

***7.** Show that it is not always possible to express a logic form by means of a logically equivalent one which uses only the logical operations $\rightarrow$, $\vee$.

[Hint: Show that you cannot build any contradictions using only the operations $\rightarrow$ and $\vee$.]

---

*This denotes a difficult problem.

3

Analysis of Inferences

3.1
Sentential Validity

You have probably met the following types of argument before. The first example constitutes good reasoning, while the second one does not.

Example 1. If grass is green, snow is white.

Grass is green.
____
Therefore, snow is white.

Example 2. If x and y are even numbers, x + y is even.

x + y is even.
____
Therefore, x and y are even numbers.

Each of these examples is called an **inference** or an **instance of an inference pattern.** An **inference pattern** is a finite list of logic forms presented as follows:

$$\begin{array}{c} \mathbf{A_1} \\ \mathbf{A_2} \\ \vdots \\ \mathbf{A_n} \\ \hline \therefore \mathbf{B} \end{array}$$

The logic forms above the bar are called **premises** while **B** is called the **conclusion.** (The symbol ∴ is read "therefore.")

An inference pattern is **sententially valid** if and only if

(conjunction of the premises) → conclusion

is a tautology. An inference is called sententially valid if its inference pattern is sententially valid.

Example 3. Consider the inference:

If grass is green, snow is white.
Grass is green.
Therefore, snow is white.

To check it for sentential validity, first symbolize.

G: Grass is green
S: Snow is white.

The inference pattern is

$$\begin{array}{l} G \rightarrow S \\ \underline{G \qquad} \\ \therefore S \end{array}$$

Since

$$[(G \rightarrow S) \wedge G] \rightarrow S$$

is a tautology (verify that yourself), the inference is sententially valid.

The sentential validity of an inference depends solely on the way in which the premises and conclusion are built from their basic sentences by means of the logical operations. You should convince yourself that if you substitute sentences for the variable symbols in a sententially valid inference pattern, and if by doing so you get premises which are true statements, then the conclusion will also be a true statement. Thus an inference which is sententially valid constitutes good reasoning.

Example 4. You observed above that

$$\begin{array}{l} A \rightarrow B \\ \underline{A \qquad} \\ \therefore B \end{array}$$

is a sententially valid inference pattern. An inference which is an instance of this pattern is:

If cars sing, dogs whistle.

Cars sing.

Therefore, dogs whistle.

This inference, then, is a sententially valid inference.

The preceding example vividly demonstrates that it is only the *form* of the inference that determines whether or not it is valid. The fact that the premises and the conclusion are unrealistic has no bearing on the matter.

Study the following important example.

Example 5. All countries have armies.

France is a country.

Therefore, France has an army.

Let A be: All countries have armies.
Let B be: France is a country.
Let C be: France has an army.

The inference pattern corresponding to the above inference is:

$$\begin{array}{l} A \\ \underline{B} \\ \therefore C \end{array}$$

It is easy to see that the logic form

$$(A \wedge B) \rightarrow C$$

is *not* a tautology. Thus the inference is *not* sententially valid. On the other hand, any reasonable person would probably agree that the inference's conclusion really does follow from its premises. The reason that a sentential validity test doesn't help to recognize that this inference constitutes correct reasoning is that its correctness is somehow related to the way in which the premises and conclusion are built from their respective subjects and predicates. B and C have the same subject while A and C have the same predicate. Our symbolic resources are not adequate for the job of revealing this. Logic forms display only the way in which compound sentences are built from their basic elementary sentences; they do not display the subject-predicate structure of the elementary sentences themselves. The problem of how to recognize the correctness of such inferences involves an area of logic known as "quantification theory," which is beyond the scope of this book and will not be pursued.

Remember: Inferences which are sententially valid constitute good reasoning. Inferences which are not sententially valid might represent good reasoning nevertheless.

Exercises 3.1

1. Test the following inference patterns for sentential validity.

(a)
$$\begin{array}{l} A \vee B \\ B \vee C \\ \hline \therefore A \end{array}$$

(b)
$$\begin{array}{l} A \rightarrow B \\ C \rightarrow A \\ \hline \therefore C \rightarrow B \end{array}$$

(c)
$$\begin{array}{l} A \rightarrow B \\ \neg B \rightarrow \neg C \\ \hline \therefore B \vee \neg C \end{array}$$

(d)
$$\begin{array}{l} A \vee B \\ \neg B \vee C \\ \hline \therefore A \vee C \end{array}$$

(e)
$$\begin{array}{l} \neg A \\ \hline \therefore \neg (A \wedge B) \end{array}$$

(f)
$$\begin{array}{l} A \rightarrow (B \rightarrow C) \\ \neg B \rightarrow \neg A \\ A \\ \hline \therefore C \end{array}$$

(g)
$$\begin{array}{l} \neg D \rightarrow \neg B \\ \neg C \\ D \rightarrow A \\ B \vee C \\ \hline \therefore A \end{array}$$

(h)
$$\begin{array}{l} \neg (B \vee C) \rightarrow \neg B \\ D \rightarrow \neg C \\ B \rightarrow \neg A \\ \hline \therefore A \rightarrow \neg D \end{array}$$

2. Determine which of the following inferences are sententially valid.

(a) If the legislature meets today, the executive committee met yesterday.

The executive committee met yesterday.

Therefore, the legislature meets today.

(b) If wages are raised, inflation continues.

If there is a depression, wages cannot be raised.

Therefore, if there is a depression, inflation cannot continue.

(c) If this child is challenged, then he will enjoy learning. If he is not challenged, this child will be bored with school.

This child does not enjoy learning.

Therefore, he is bored with school.

(d) If other countries develop new weapons, we feel that our national security is threatened. If we do not feel that our national security is threatened, then we will spend more money on social reform programs.

Therefore, if other countries develop new weapons, then we will not spend more money on social reform programs.

(e) If the Department of Agriculture is correct, corn blight does not occur if the crop is sprayed weekly.

Corn blight does occur.

Therefore, if the corn is sprayed weekly, the Department of Agriculture is wrong.

(f) Unless protective tariffs are imposed, our balance of payments will be unfavorable. If our balance of payments is unfavorable, the foreign aid appropriations will be cut.

Therefore, if protective tariffs are imposed, foreign aid appropriations will not be cut.

3. Give an example of an inference which is not sententially valid, but which has a true conclusion.

4. Give an example of an inference which is sententially valid, but which has a false conclusion.

5. Give an example of an inference which is sententially valid and which has false premises and a false conclusion.

3.2 Basic Inferences

Here is a list of three basic inference patterns and their names. Each of these can be shown to be sententially valid by the method introduced in the preceding section. These three inferences will be so essential in later work that you are urged to memorize them.

I. A

$$\frac{A \rightarrow B}{\therefore B}$$ **detachment**

II. $A \rightarrow B$

$$\frac{B \rightarrow C}{\therefore A \rightarrow C}$$ **hypothetical syllogism**

III. $A \rightarrow C$

$$\frac{B \rightarrow C}{\therefore (A \vee B) \rightarrow C}$$ **case inference**

Exercises 3.2

1. Verify that the three inference patterns just given are sententially valid.

3.3
Checking Sentential Validity of Inferences by Repeated Use of Previously Proven Inferences

The method introduced earlier for checking the sentential validity of an inference has the advantage of being conclusive in that it ultimately leads to an answer. Unfortunately, it sometimes leads to involved symbolic manipulations. The technique described now seems closer to the natural reasoning process and is often shorter. However, this new method is recommended only if you already suspect that the inference in question is sententially valid.

In the new technique, inferences that have already been shown to be sententially valid (in particular those introduced in the preceding section) are used to prove that the given inference is sententially valid. The method is based on the following rule:

> An inference is sententially valid if the conclusion appears as the last line in a vertical list of logic forms,

each of which satisfies one of the following four conditions:

1. it is a premise,
2. it is a tautology,
3. it is logically equivalent to a logic form preceding it in the list,
4. it follows from logic forms preceding it in the list by virtue of a previously proven inference.

Example 1. Verify:

$$\begin{array}{l} A \rightarrow B \\ \neg B \\ \hline \therefore \neg A \end{array}$$

DEMONSTRATION:

(1)	$A \rightarrow B$	premise
(2)	$\neg B \rightarrow \neg A$	equivalent to (1) by the law of the contrapositive
(3)	$\neg B$	premise
(4)	$\neg A$	Detachment; 3,2.

Example 2. Verify:

$$\begin{array}{l} A \rightarrow B \\ C \rightarrow D \\ A \vee C \\ \hline \therefore B \vee D \end{array}$$

DEMONSTRATION:

(1)	$A \rightarrow B$	premise
(2)	$B \rightarrow (B \vee D)$	tautology
(3)	$A \rightarrow (B \vee D)$	Hypothetical Syllogism; 1,2
(4)	$C \rightarrow D$	premise
(5)	$D \rightarrow (B \vee D)$	tautology
(6)	$C \rightarrow (B \vee D)$	Hypothetical Syllogism; 4,5
(7)	$(A \vee C) \rightarrow (B \vee D)$	Case Inference; 3,6
(8)	$A \vee C$	premise
(9)	$B \vee D$	Detachment; 8,7

A chain of inferences like the one in the preceding example is called a **deduction** of the conclusion from the premises. Such deductions resemble mathematical proofs (for example, the proofs of Greek geometry). This is not to say, however, that all proofs in geometry are such simple deductions. Analysis of the reasoning used in geometry has shown that many of the inferences used there depend not only on the way in which the premises and conclusions are built from their elementary sentences by means of the logical operations, but also on the internal (subject—predicate) structure of the elementary sentences as well. The ancient Greeks did not carry the analysis of these inferences far enough; the logical analysis of the Greeks was inadequate for making fully explicit the reasoning used in their own mathematics.

Exercises 3.3

1. Verify that the following inferences are valid by giving a deduction from the premises to the conclusion. (The list of basic equivalences in section 2.2 should be very helpful.)

(a)
$$\begin{array}{l} \quad A \\ \hline \therefore B \rightarrow A \end{array}$$

(b)
$$\begin{array}{l} A \rightarrow B \\ \neg C \rightarrow \neg B \\ C \rightarrow D \\ \hline \therefore A \rightarrow D \end{array}$$

(c)
$$\begin{array}{l} A \rightarrow B \\ C \rightarrow \neg B \\ C \\ \hline \therefore \quad A \end{array}$$

(d)
$$\begin{array}{l} (A \rightarrow B) \vee (A \rightarrow C) \\ A \\ \hline \therefore B \vee C \end{array}$$

(e)
$$\begin{array}{l} A \rightarrow B \\ A \vee B \\ \hline \therefore B \end{array}$$

(f)
$$\begin{array}{l} B \rightarrow A \\ D \rightarrow C \\ B \vee D \\ \hline \therefore A \vee C \end{array}$$

(g)
$$\begin{array}{l} \neg A \rightarrow (B \rightarrow \neg C) \\ \neg A \vee \neg D \\ \neg E \rightarrow B \\ \quad\quad D \\ \hline \therefore C \rightarrow E \end{array}$$

4

Switching Circuits

4.1
Representing Switching Circuits By Logic Forms

In this chapter the relationship between the design of switching circuits and the algebra of logic forms will be investigated. First you will see how to represent a switching circuit by means of a logic form. This is done by making use of some basic properties of electric circuits and switches.

To each switching circuit you will have to associate a **circuit diagram.** Examples of such diagrams are:

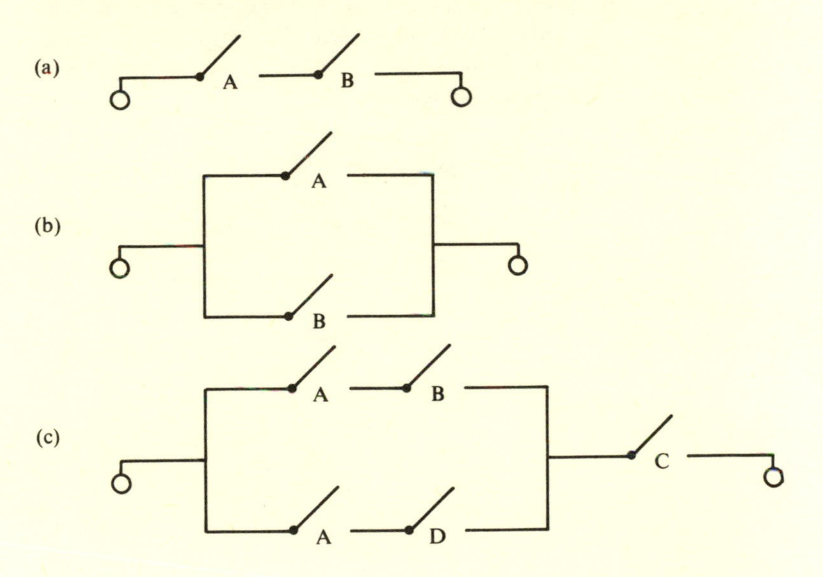

The symbol “ ” represents a terminal of the circuit, while the symbol “ ” represents a switch in the circuit.

You know that for current to flow from one terminal to the other, there must be an unbroken path connecting the terminals. A path is unbroken if and only if all the switches occurring along it are closed. Using this fact, you can assign to each circuit a logic form which reflects the structure of the circuit. The following examples demonstrate this.

Example 1. Consider the circuit

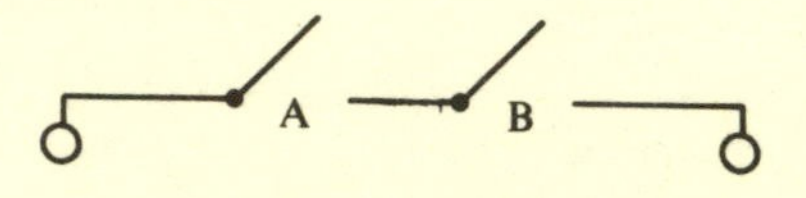

Switches joined together in this way are said to be connected in **series.** Assign the logic form A ∧ B to this circuit because current flows if and only if both switches A *and* B are closed.

Example 2. The following two switches are said to be connected in **parallel.**

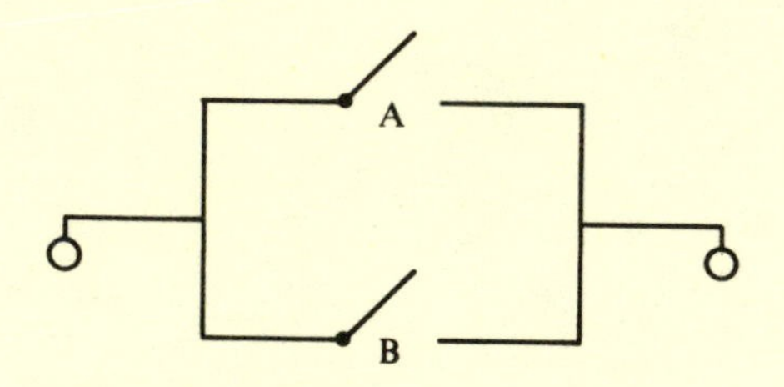

Assign the logic form A ∨ B to this circuit because current flows if and only if A is closed *or* B is closed.

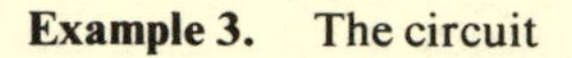

Example 3. The circuit

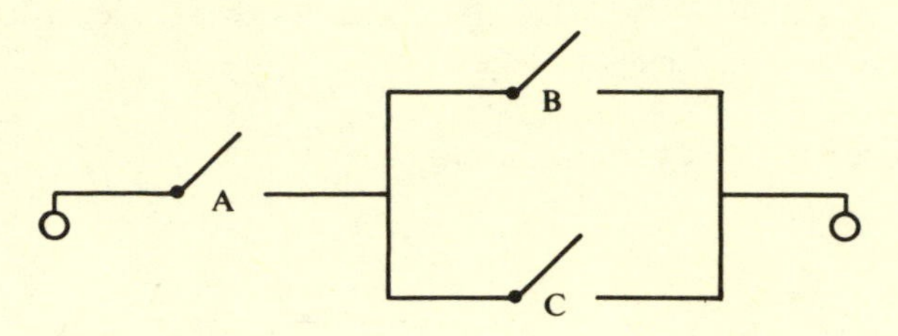

is an example of a series-parallel combination. Its logic form is A ∧ (B ∨ C) because current flows if and only if

A is closed,

and

B or C is closed.

In some circuits, one handle operates several switches simultaneously. Whenever one handle closes or opens several switches together, these switches are labeled with the same letter.

Example 4. In the circuit

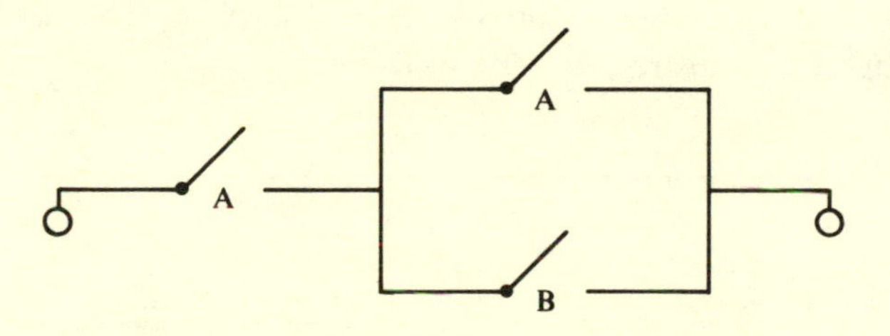

both switches labeled A act in concert and are operated by the same handle. Its associated logic form is

$$A \wedge (A \vee B).$$

This circuit has three switches, but only two control handles, an A-handle and a B-handle.

If one handle simultaneously opens one switch while closing another, it is indicated by labeling the one switch by a letter and the other by the negation of that letter.

Example 5. In the circuit

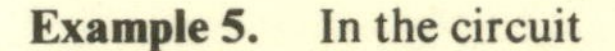

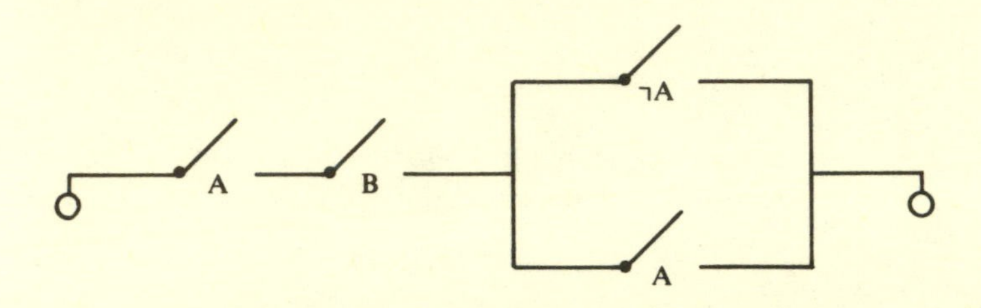

the two switches labeled A act in concert but in opposition to the switch labeled ¬ A. All three are controlled by one common handle. This circuit therefore, has four switches and two control handles, an A-handle and a B-handle. Its logic form is

$$A \wedge B \wedge (\neg A \vee A).$$

The following table summarizes the procedure used to determine the logic form which represents a switching circuit.

Diagram Representation	Type of Connection	Associated Logic Form	Reason
A B (switches in series)	series	$A \wedge B$	Both A and B must be closed for current to flow
A, B (switches in parallel)	parallel	$A \vee B$	A or B must be closed for current to flow

Exercises 4.1

1. Write the logic form associated with each of the following switching circuits.

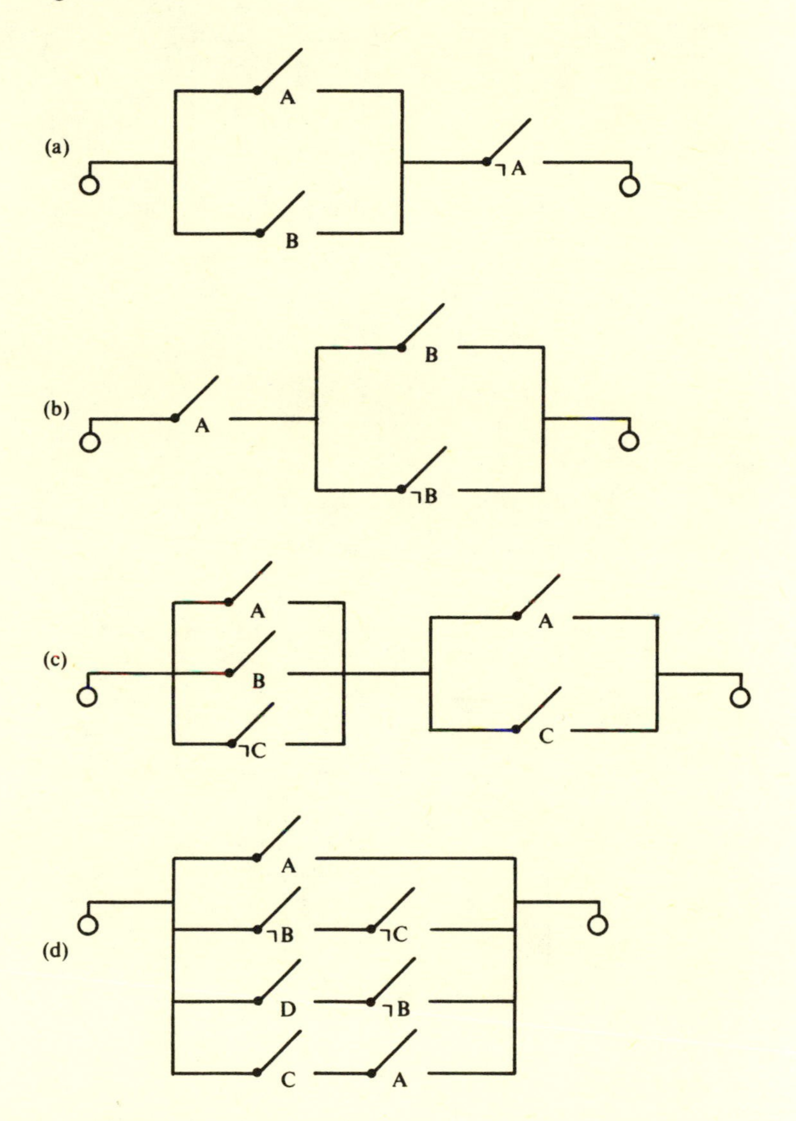

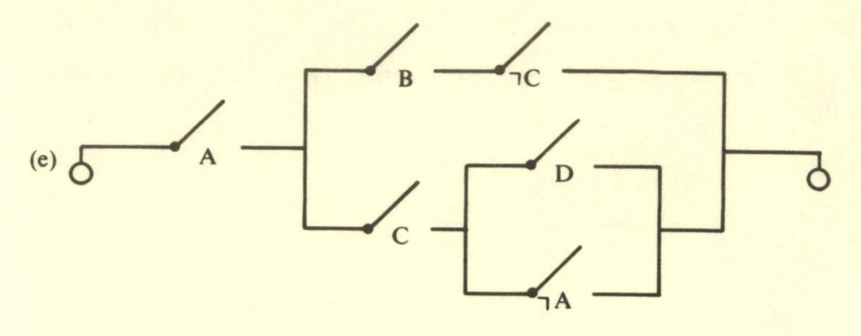

2. Draw the circuit associated with each of the following logic forms:

(a) $A \wedge [(A \wedge B) \vee B \vee (\neg A \wedge \neg B)]$
(b) $(A \wedge B) \vee [(C \vee A) \wedge \neg B]$
(c) $A \vee (B \wedge \neg A) \vee (C \wedge \neg B \wedge A) \vee \neg C$
(d) $E \wedge [F \vee G \vee A] \wedge [F \vee (\neg G \wedge \neg E)]$

4.2
Simplifying Switching Circuits

Consider the following switching circuits:

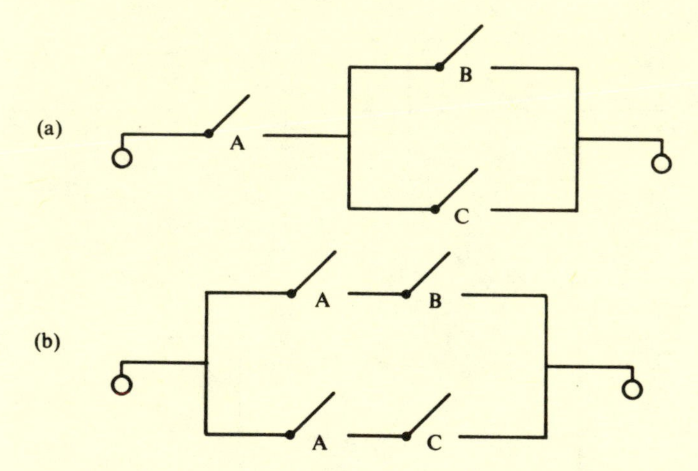

For a given setting of the handles, current flows in the first circuit if and only if it flows in the second circuit. The first one, however, has fewer switches. Thus, the first circuit may be called a **simplification** of the second. Actually there is a very practical reason for wanting to simplify a switching circuit; the smaller the number of switches in a circuit, the lower the cost of manufacturing it. Imagine how much money the telephone company could save if it eliminated just one switch from each bank of switches in the phone system!

In Chapter 2 you learned how to simplify a logic form by algebraic manipulations. Now you can use the skills you developed there to simplify switching circuits by doing the following:

1. Determine the logic form associated with the circuit.
2. Use algebraic manipulations to simplify the logic form.
3. Determine the switching circuit associated with the simplified logic form.

Example. Simplify the following switching circuit and sketch the simplified circuit.

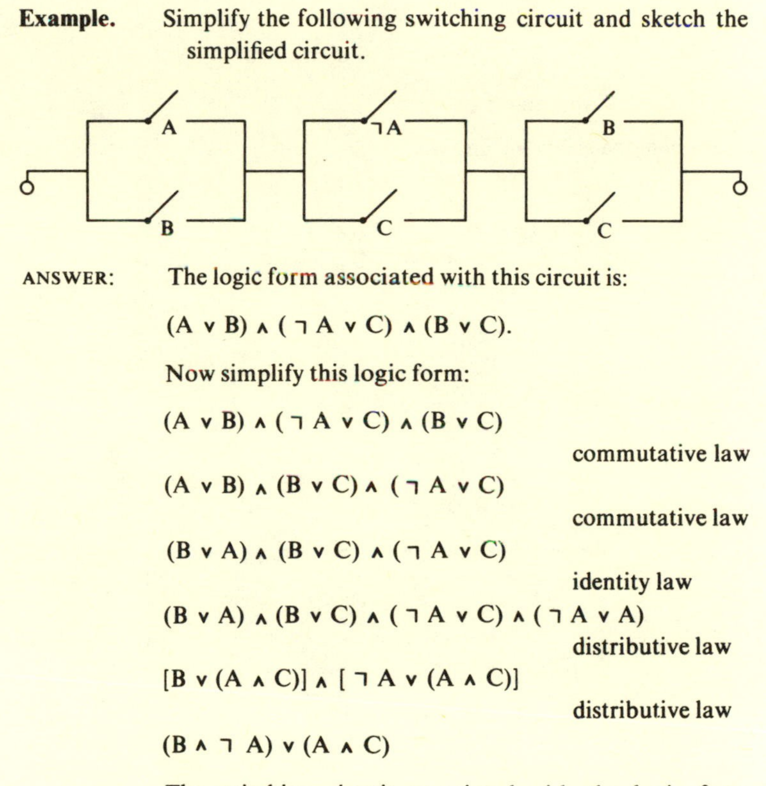

ANSWER: The logic form associated with this circuit is:

$(A \vee B) \wedge (\neg A \vee C) \wedge (B \vee C)$.

Now simplify this logic form:

$(A \vee B) \wedge (\neg A \vee C) \wedge (B \vee C)$

commutative law

$(A \vee B) \wedge (B \vee C) \wedge (\neg A \vee C)$

commutative law

$(B \vee A) \wedge (B \vee C) \wedge (\neg A \vee C)$

identity law

$(B \vee A) \wedge (B \vee C) \wedge (\neg A \vee C) \wedge (\neg A \vee A)$

distributive law

$[B \vee (A \wedge C)] \wedge [\neg A \vee (A \wedge C)]$

distributive law

$(B \wedge \neg A) \vee (A \wedge C)$

The switching circuit associated with the logic form

$(B \wedge \neg A) \vee (A \wedge C)$ is:

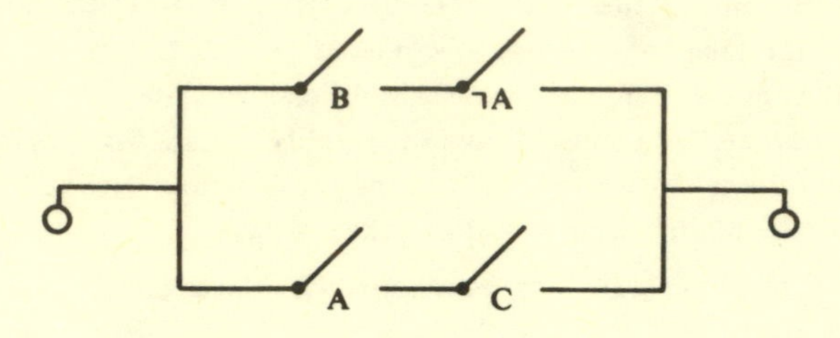

This circuit is a simplification of the circuit with which you started.

Exercises 4.2

1. Simplify the following circuits.

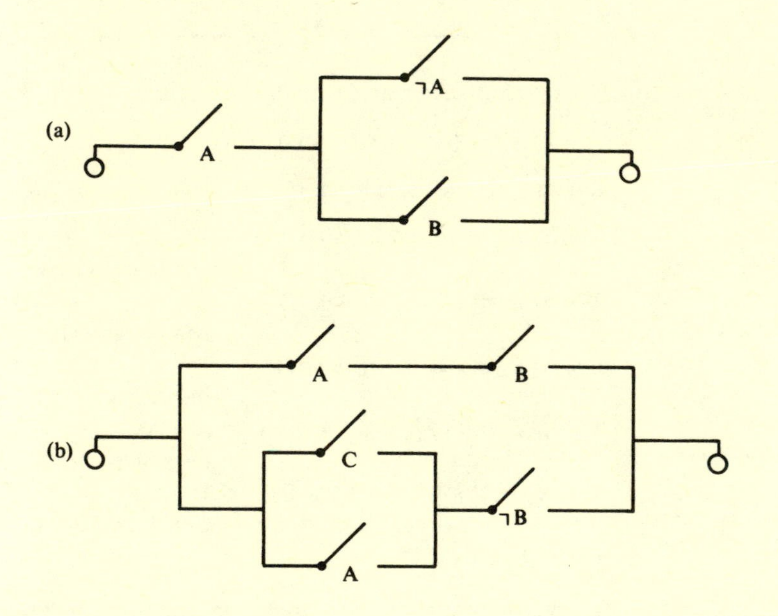

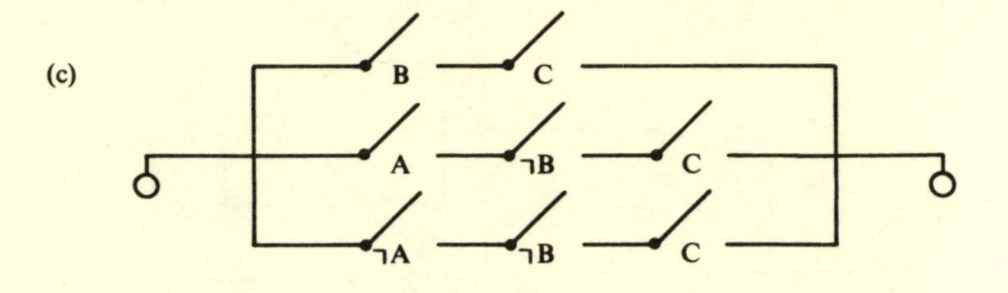
(c)
B
C
A
¬B
C
¬A
¬B
C

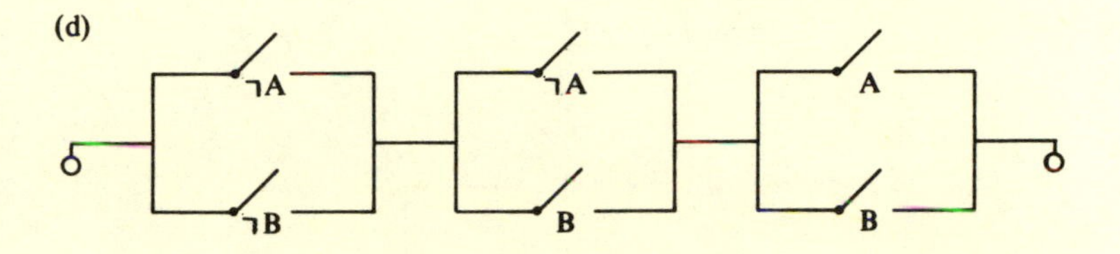
(d)
¬A
¬A
A
¬B
B
B

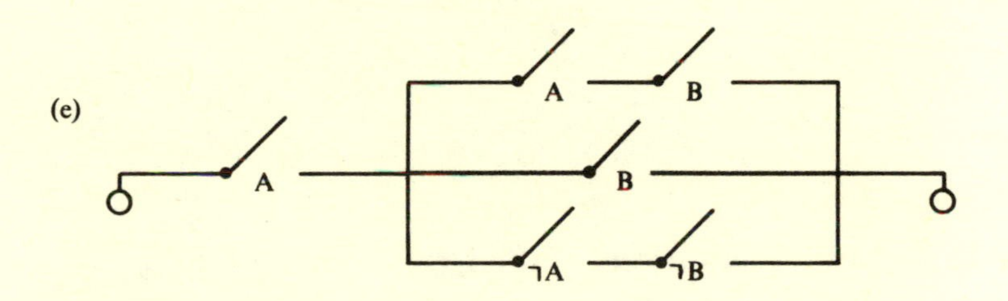
(e)
A
B
A
B
¬A
¬B

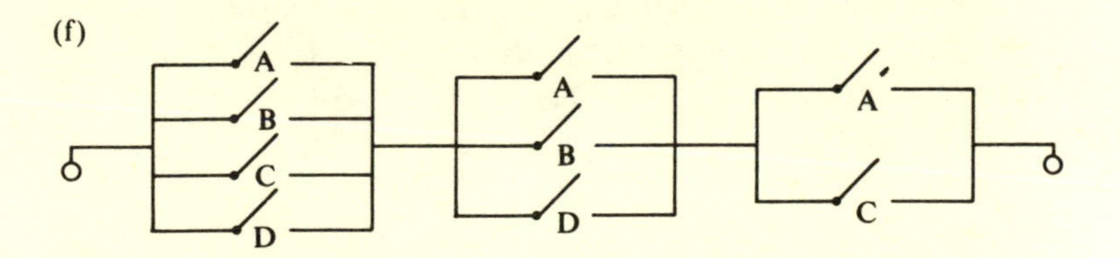
(f)
A
B
C
D
A
B
D
A
C

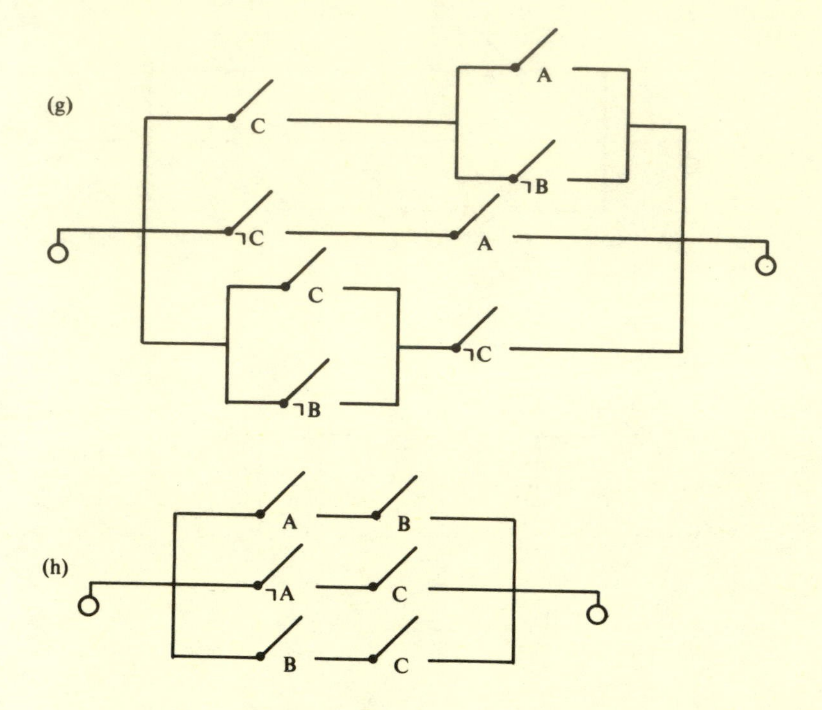

4.3
Limitations of the Logic Form Method Of Simplifying Switching Circuits

In the last section logic forms were used to simplify certain circuits. There are two main drawbacks to this method. First, there is not yet a "nice" practical way to determine if a switching circuit is simplified as much as possible. In fact, this problem is still being investigated by research mathematicians. Also, this method does not take into account circuits which are not "series-parallel" circuits.

Example. The following is a diagram of a "bridge" circuit.

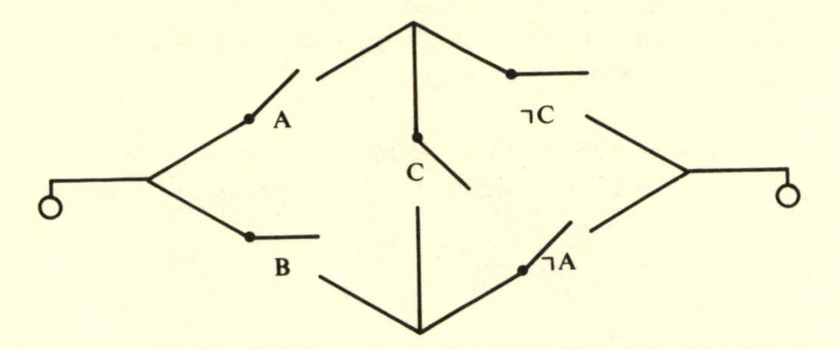

The circuits which correspond to logic forms are "series-parallel" circuits, not "bridge" circuits. Thus, our method would never lead us to a circuit like this. Consequently, if this were the simplest circuit for a given situation, our method would fail to reveal that fact.

Exercises 4.3

1. Find a "series-parallel" circuit which behaves like the following "bridge" circuit.

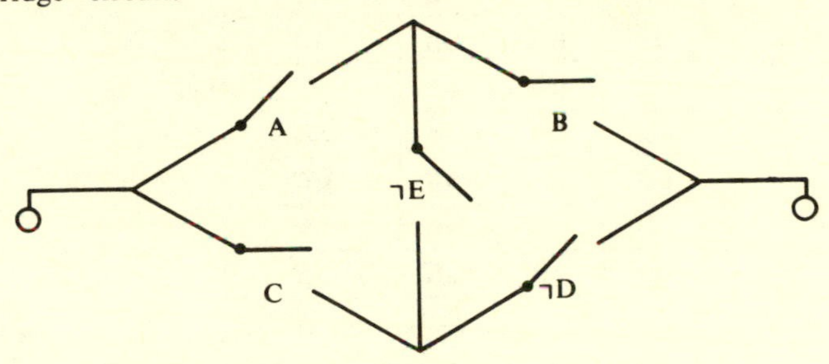

4.4
Designing a Switching Circuit Starting With a Table Which Describes Its Behavior

By examining the diagram of any switching circuit, you can immediately determine whether current flows for any given setting of the switches. Then you can compile a table that shows what the circuit will do for each setting of the switches.

Example. Display a table describing the behavior of the following switching circuit for each possible setting of its switches.

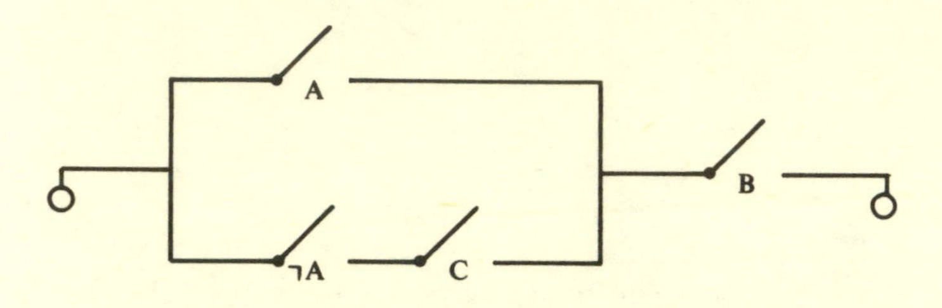

ANSWER: The table should appear as follows:

A	B	C	CIRCUIT
on	on	on	on
on	on	off	on
on	off	on	off
on	off	off	off
off	on	on	on
off	on	off	off
off	off	on	off
off	off	off	off

Now suppose you are given a table like the one constructed in the preceding example. It may seem surprising, but by using techniques developed earlier, you can actually design a switching circuit whose behavior is prescribed by the given table! First think of the table as a truth table (with "on" playing the role of the truth value t). Then construct the logic form which has this table for its truth table (You learned how to do this in Section 1.8.) After that, construct the circuit which corresponds to this logic form. This will be the desired circuit.

Example 1. Design a switching circuit whose behavior is described by the following table, and sketch this circuit.

A	B	CIRCUIT	
on	on	on	√
on	off	off	
off	on	off	
off	off	on	√

ANSWER: 1. Restrict your attention to the rows with checks: row 1 and row 4.

2. The sequence of terms corresponding to row 1 is

$$A, B,$$

and the resulting conjunction is

$$A \wedge B.$$

The sequence of terms corresponding to row 4 is

$$\neg A, \neg B,$$

and the resulting conjunction is

$$\neg A \wedge \neg A.$$

3. The disjunction you are seeking is

$$(A \wedge B) \vee (\neg A \wedge \neg B)$$

and the diagram of the corresponding switching circuit is

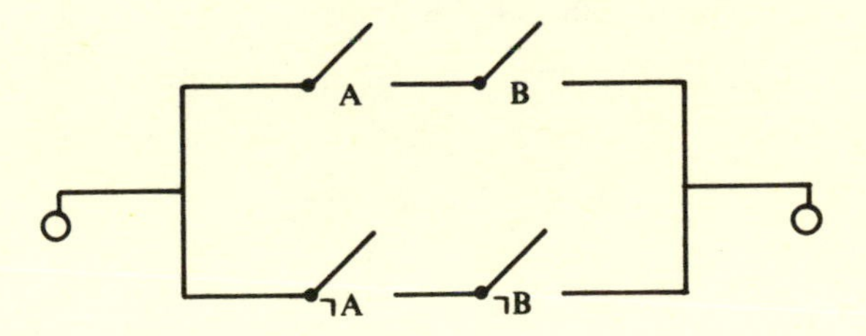

The method just outlined will always lead to a switching circuit with the desired behavior. However, the circuit obtained by this method may not be the simplest circuit with this behavior.

Example 2. Design and simplify a circuit whose behavior is described by the table:

A	B	CIRCUIT
on	on	on
on	off	off
off	on	on
off	off	off

ANSWER: The method just learned shows that the table describes the behavior of the switching circuit corresponding to the logic form:

$(A \wedge B) \vee (\neg A \wedge B)$

Now this can be simplified.

$(A \wedge B) \vee (\neg A \wedge B)$

distributive law

$(A \vee \neg A) \wedge B$

identity law

B

Hence the table also describes the behavior of the very simple circuit:

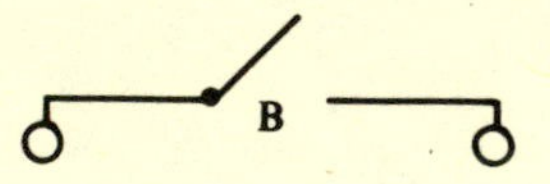

Exercises 4.4

1. Design and simplify a circuit whose behavior is described by the following table:

A	B	C	CIRCUIT
on	on	on	on
on	on	off	off
on	off	on	off
on	off	off	off
off	on	on	on
off	on	off	off
off	off	on	off
off	off	off	off

2. Construct a circuit to control a light switch from three different locations. [Hint: Make a table to describe the behavior of the circuit. From the table, construct the logic form and then the circuit.]

3. The five members of the Pookaville city countil are meeting to vote on a new tax bill. Each man votes in secret by flipping a switch. Design a circuit so that a light will come on if and only if a majority of the council votes "yes" for the bill. [Hint: Make a table.]

4. A certain king's five advisors must decide whether to ratify a proposed treaty with the United States. It takes a majority vote of the advisors to ratify the treaty, except that the advisor who is prime minister has a veto (i.e., the treaty is ratified only if he votes for it). Design a circuit so that each advisor can vote for ratification by throwing a switch, and a bell rings if and only if the treaty is ratified. [Hint: Make a table.]

5

Set Theory

5.1
Elementary Set Theory

The idea of a **set** is the foundation upon which most of modern mathematics is built. It is so fundamental that we will not even try to define it. Synonyms like "collection" or "class" or "family" are often used to convey the same idea.

The objects which make up a set are called the **elements** or **members** of the set. If an object x is an element of the set A, we write

$$x \in A.$$

If x is not an element of A, we write

$$x \notin A.$$

If every member of A is also a member of B, we say A is a **subset** of B, and write

$$A \subseteq B.$$

It is easy to see that every set is a subset of itself. Two sets, A and B, are **equal** (written $A = B$) if and only if they have the same members; that is, every member of A is a member of B and vice versa. In case $A \subseteq B$, but $A \neq B$, we say A is a **proper subset** of B and write

$$A \subset B.$$

This happens when every member of A is a member of B, but some member of B is not a member of A.

Example 1. The box on the next page should help to clarify the notions introduced so far.

SET	ONE OF ITS MEMBERS	ONE OF ITS SUBSETS
The set of all people	Ringo Starr	The set of all women
The set of teachers at Dogwood High	Joe Dodo	The set of math teachers at Dogwood High
The set of all whole numbers	3	The set of all even whole numbers
The Peanuts family	Snoopy	The set consisting of Snoopy and Lucy
The set of all airplanes	The Spirit of St. Louis	The set of all jet planes

In many applications of the theory of sets all the sets which come into play are built from elements taken from a set U which is specified in advance and called the **universe** (for that application). For example, in plane geometry the universe U is usually taken to be the set of all points in the plane, and the geometric figures like lines, circles, and triangles are thought of as sets of such points.

In denoting sets, **set braces** { , } are often used. A set is sometimes specified by listing its elements inside set braces. For example, {1, 2, 8} is the set whose elements are 1, 2, and 8. In specifying a set this way, it is immaterial in what order the elements are written and how often they are repeated. Thus

$$\{1, 2\} = \{2, 1\} = \{2, 2, 1\}$$

A set can also be specified by stating a property its members will have. For example,

> "the set of all objects which have the property of being an odd number,"

in short,

> "the set of all odd numbers."

Using x to denote a typical member of this set, we write

$$\{x \mid x \text{ is an odd number}\}$$

and read this

"the set of all x such that x is an odd number."

The symbol " | " is read "such that."

It is convenient to introduce the so-called **empty set** which has no members. The symbol $\emptyset$ is used to denote it. It can easily be shown that $\emptyset$ is a subset of any set A by the following reasoning. To assert $\emptyset \subseteq A$ is the same as asserting that for any x, the implication

$$x \in \emptyset \rightarrow x \in A$$

holds; but this implication holds automatically since for any x, the left side of the implication is false.

If A and B are sets, the set of elements belonging to either A or B is called the **union** of A and B. We write this

$$A \cup B.$$

For example, $\{1, 2\} \cup \{2, 3\} = \{1, 2, 3\}$.

The set of elements in both A and B is called the **intersection** of A and B. We write this

$$A \cap B.$$

For example, $\{1, 2\} \cap \{2, 3\} = \{2\}$.

The set of elements which is in A but not in B is called the **difference** of A and B. We write this

$$A - B.$$

For example, $\{1, 2\} - \{2, 3\} = \{1\}$.

When a universe U is specified and A is a subset of U, U–A is called the **complement** of A (with respect to the universe U). The complement of A is denoted by

$$A'.$$

For example, if U is the set of all whole numbers, and A is the set of all even numbers, then A′ is the set of all odd numbers.

Example 2. Suppose universe U is the set of all positive whole numbers.

Let $A = \{1, 2, 3, 4, 5\}$
Let $B = \{1, 3, 4, 9\}$
Let $C = \{6, 7\}$

Then $A \cup B = \{1, 2, 3, 4, 5, 9\}$
$A \cap B = [1, 3, 4\}$
$A \cap C = \emptyset$
$A - B = \{2, 5\}$
$A' = \{6, 7, 8, 9 \ldots\}$

It is possible for the elements of a set to be sets themselves. For example, for each set A, the **power set** of A can be formed which has the subsets of A as its elements. It can be shown that if A has n elements, the power set of A has 2^n elements.

Example 3. If $A = \{1, 2, 3\}$, the elements of the power set of A are:

$\emptyset$, $\{1\}$, $\{1, 2\}$, $\{1, 2, 3\}$

$\{2\}$, $\{3\}$, $\{2, 3\}$, $\{1, 3\}$

Relationships among sets can be visualized by making a type of picture called a **Venn diagram.** A few diagrams are given below. Think of the areas enclosed by the rectangles as the universe U. The rest ought to be self-explanatory.

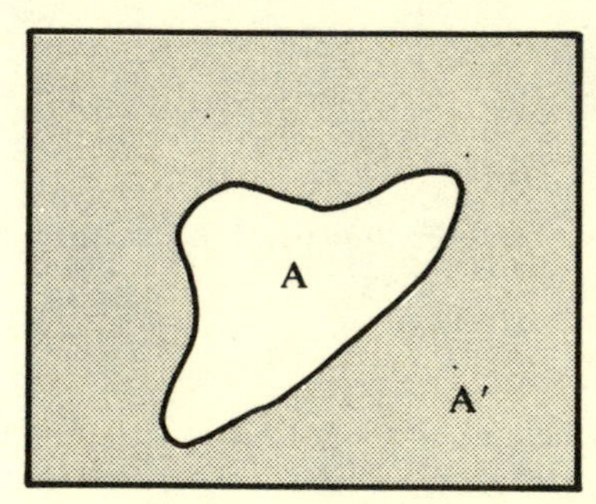

A and its complement A′

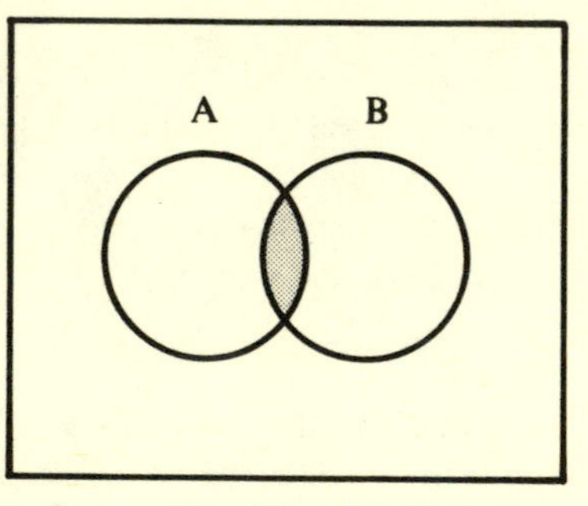

Intersection of A and B (shaded)

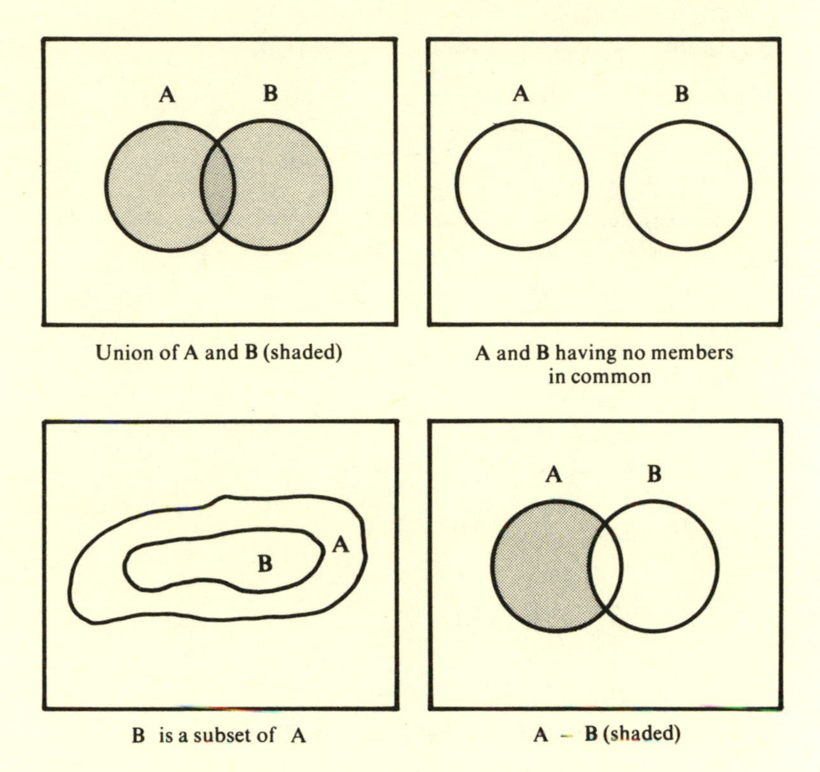

Union of A and B (shaded)

A and B having no members in common

B is a subset of A

A − B (shaded)

Exercises 5.1

1. Are the sets {1, 1, 1} and {1} equal? Why?

2. Let U = {a, b, c, d, e, x, y, z}
   Let A = {a, b, c}
   Let B = {a, e, x}
   Let C = {x, y, z}

   It is also given that the objects denoted by the small letters are all different. Exhibit the following sets.

   (a) A' (b) $A \cup B$
   (c) $A - C$ (d) $A \cap C$
   (e) $(A \cap C)'$ (f) $A \cup B \cup C$
   (g) $B - C$ (h) $C - B$

3. List the elements in the power set of $A = \{a, b, c, d\}$.

4. Given: $A = \{1, b, c\}$
$B = \{e, f, b, g, h\}$
$C = \{1, 2, c, b\}$
$D = \{a, c, e, i\}$

Assuming that all of the symbols in the set braces above denote distinct objects, determine which of the following are true:

(a) $c \in [(B \cup A) \cap C]$
(b) $b \in (A \cap B \cap C)$
(c) $e \in (A \cap D)$
(d) $2 \in (C \cup D)$
(e) $(C \cap D) \subset (C \cap A)$
(f) $(A \cup C) = C$
(g) $B \cap (D \cap A) \neq \emptyset$
(h) $(C \cap D) \cap B \neq \emptyset$
(i) $(A \cap B) \cup (C \cap D) = \{b, c\}$
(j) $(A \cup D) \cap C \neq A$
(k) $C \cap D = \emptyset$
(l) $(B \cap A) \cap (B \cap C) = \emptyset$

5.2 Basic Identities and Relations of Set Theory

In this section we have listed the set identities and relations most frequently needed when using set theory to solve mathematical problems. It is assumed that all the sets appearing in the following list are subsets of a given universe U.

Basic Identities

Identity	Law
1. $(A')' = A$	law of double complement
2. $A \cap B = B \cap A$ 3. $A \cup B = B \cup A$	commutative laws
4. $(A \cup B) \cup C = A \cup (B \cup C)$ 5. $(A \cap B) \cap C = A \cap (B \cap C)$	associative laws

6.	$A \cap (B \cup C) = (A \cap B) \cup (A \cap C)$	distributive laws
7.	$A \cup (B \cap C) = (A \cup B) \cap (A \cup C)$	
8.	$(A \cup B)' = A' \cap B'$	de Morgan's laws
9.	$(A \cap B)' = A' \cup B'$	
10.	$A \cap A = A$	idempotent laws
11.	$A \cup A = A$	
12.	$A \cap (A \cup B) = A$	absorption laws
13.	$A \cup (A \cap B) = A$	
14.	$A \cap U = A$	identity laws
15.	$A \cup \phi = A$	
16.	$A \cup U = U$	domination laws
17.	$A \cap \phi = \phi$	
18.	$A \subseteq B$ if and only if $A' \cup B = U$	
19.	$A \subseteq B$ if and only if $B' \subseteq A'$	
20.	$A = B$ if and only if $A \subseteq B$ and $B \subseteq A$.	

You have, no doubt, been struck by the formal similarity between these identities and the basic logical equivalences used to manipulate logic forms. More will be said about this later.

In the example below, one of these identities will be verified for you. The rest of them can be done in a similar way.

Example. Verify de Morgan's law (for sets).

$$(A \cup B)' = A' \cap B'$$

PROOF: Let x denote an arbitrary object; then,

$x \in (A \cup B)' \equiv \neg [x \in (A \cup B)]$ definition of complement

$\equiv \neg [(x \in A) \vee (x \in B)]$ definition of union

$\equiv \neg (x \in A) \wedge \neg (x \in B)$ de Morgan's law (for logic)

$\equiv (x \in A') \wedge (x \in B')$ definition of complement

$\equiv x \in (A' \cap B')$ definition of intersection

thus $x \in (A \cup B)' \equiv x \in (A' \cap B')$

Therefore $(A \cup B)' = A' \cap B'$

Exercises 5.2

All the sets appearing in these exercises are subsets of a given universe U.

1. Under what circumstances are the following statements true?

(a) $A \cap B' = A$
(b) $A \subseteq \emptyset$
(c) $A \cup B = A \cap B$
(d) $A - B = B - A$
(e) $A \cap B = A$
(f) $(A \cup B) \cap B' = A$
(g) $(A \cap B') \cup B = A \cup B$

2. Verify the following identities.

(a) $(A')' = A$
(b) $A \cap (B \cup C) = (A \cap B) \cup (A \cap C)$
(c) $A \cap (A \cup B) = A$
(d) $(A \cap B)' = A' \cup B'$
(e) $(A \cup B) - (A \cap B) = (A - B) \cup (B - A)$.
(f) $A = (A - B) \cup (A \cap B)$

5,3
Checking Consistency of Data

Social scientists working for the government spend much time collecting and evaluating data to reach conclusions that help shape government programs. Private enterprise spends much money on market research in which data is collected and evaluated to determine the public's response to a company's products. It may surprise you to learn that some very elementary set theory can be useful in checking that the collected data make sense, but this is indeed the case!

If A is a set, let $\#(A)$ denote the number of elements in A. The following formula, although easy to verify, is important.

$$\#(A \cup B) = \#(A) + \#(B) - \#(A \cap B)$$

Using it, you can now derive a similar formula for $\#(A \cup B \cup C)$. The basic idea is to think of $A \cup B \cup C$ as the union of two sets, namely A and $(B \cup C)$. Now apply the above formula to these two sets to get

$$\#(A \cup [B \cup C]) = \#(A) + \#(B \cup C) - \#(A \cap [B \cup C]).$$

The formula can further be applied to the second term. Moreover, after applying the distributive law to the third term, the formula can be applied to its parts. Collecting terms, the result is

$$\begin{aligned} \#(A \cup B \cup C) = \#(A) + \#(B) + \#(C) \\ - \#(A \cap B) - \#(A \cap C) - \#(B \cap C) \\ + \#(A \cap B \cap C) \end{aligned}$$

Study the two boxed equations! Compare them! If you look at both of them the right way, you should be able to make a good guess as to how the formula for *four* sets ought to look.

In the following two examples you will see how to use the above formulae in a practical way.

Example 1. In a random survey of 100 people who owned houses or cars, it was found that 70 people owned a house, and 30 people owned both a house and a car. How many people surveyed owned a car?

ANSWER: Let H = set of house owners
C = set of car owners

The basic formula is

$$\#(H \cup C) = \#(H) + \#(C) - \#(H \cap C).$$

Substituting the data, the result is

$$100 = 70 + \#(C) - 30.$$

Thus solving for #(C), the answer is

$\#(C) = 60.$

Hence, 60 of the people surveyed own cars.

Example 2. Data concerning the sex, education, and marital status of the 500 employees of the Sunshine Brewery were reported as follows:

250 females
156 college graduates
235 married
21 female college graduates
73 married females
43 married college graduates
12 married female college graduates

Now the man in charge of assembling the data was a bit of a boozer, and it was rumored that he made up the data after having too much to drink. His boss became suspicious and was able to check the data for consistency as follows:

F = set of female employees
G = set of college graduate employees
M = set of married employees

The data can now be written

$$\begin{aligned} 250 &= \#(F) \\ 156 &= \#(G) \\ 235 &= \#(M) \\ 21 &= \#(F \cap G) \\ 73 &= \#(M \cap F) \\ 43 &= \#(M \cap G) \\ 12 &= \#(M \cap F \cap G) \end{aligned}$$

Putting this data into the previously given formula for the union of three sets, we get

$$\begin{aligned} \#(M \cup F \cup G) &= 250 + 156 + 235 - 21 \\ &\quad - 73 - 43 + 12 \\ &= 516 \end{aligned}$$

Recall that the company had only 500 employees and $M \cup F \cup G$ being a subset of the set of employees can have no more than 500 members. This contradicts the figure of 516 just calculated, so the data is inconsistent. The data collector, by the way, was transferred to the tasting department where he seems to be getting along well.

Exercises 5.3

***1.** For the finite sets A, B, C, D, derive the formula:

$$
\begin{aligned}
&\#(A \cup B \cup C \cup D) \\
&\quad = \#(A) + \#(B) + \#(C) + \#(D) - \#(A \cap B) \\
&\quad - \#(A \cap C) - \#(A \cap D) - \#(B \cap C) \\
&\quad - \#(B \cap D) - \#(C \cap D) + \#(A \cap B \cap C) \\
&\quad + \#(A \cap B \cap D) + \#(A \cap C \cap D) + \#(B \cap C \cap D) \\
&\quad - \#(A \cap B \cap C \cap D)
\end{aligned}
$$

2. In a survey of 100 welfare recipients, the following facts were discovered:

45 had severe physical handicaps
14 were illiterate
86 lived in substandard housing
5 were illiterate and had severe physical handicaps
39 lived in substandard housing and had severe physical handicaps
12 were illiterate and lived in substandard housing
4 were illiterate, had severe physical handicaps, and lived in substandard housing.

How many of these recipients had severe physical handicaps and lived in substandard housing, but were not illiterate?

How many were illiterate but were not physically handicapped and did not live in substandard housing?

How many of those surveyed had none of the above mentioned problems?

*This denotes a difficult problem.

3. In a certain city, drivers lose their licenses if and only if they are guilty of speeding, failing to stop for a red light, or passing another car in a "no passing" zone. In 1952,

60% of those with revoked licenses were guilty of speeding,
75% failed to stop for a red light,
70% passed another car in a "no passing" zone,
45% were speeding and did not stop for a red light,
50% were speeding and passed another car in a "no passing" zone,
40% did not stop for a red light and passed another car in a "no passing" zone.

What percentage of all the drivers who lost their licenses were guilty of all three violations?

4. The United States Senate is considering three proposals to curb inflation: raising income taxes, imposing wage and price controls, and imposing higher tariffs on imports. Of the 100 senators:

62 support higher tariffs on imports
58 support higher income taxes
24 support wage and price controls

No senator supporting higher tariffs on imports favors wage and price controls. Of the senators supporting higher taxes, as many favor wage and price controls as favor higher tariffs. Moreover, each senator favors at least one of the three proposals.

(a) How many senators favor both higher taxes and wage and price controls?
(b) How many senators favor only higher taxes? Only higher tariffs? Only wage and price controls?

6

Boolean Algebra

6.1

The Abstract Approach

Earlier you were asked to learn a long list of basic logical equivalences, which were established by means of truth tables. Some of the equivalences in the list could have been derived from others listed there. This fact was not stressed at the time and now needs to be presented.

Example. The derivation below shows that the absorption law can be obtained if the commutative, identity, domination, and distributive laws hold.

$$
\begin{aligned}
\mathbf{A} \wedge (\mathbf{A} \vee \mathbf{B}) &\equiv (\mathbf{A} \vee \mathbf{B}) \wedge \mathbf{A} && \text{commutative law} \\
& && \text{identity law} \\
&\equiv (\mathbf{A} \vee \mathbf{B}) \wedge (\mathbf{A} \vee \mathbf{F}) \\
& && \text{distributive law} \\
&\equiv \mathbf{A} \vee (\mathbf{B} \wedge \mathbf{F}) \\
& && \text{domination law} \\
&\equiv \mathbf{A} \vee \mathbf{F} \\
& && \text{identity law} \\
&\equiv \mathbf{A}
\end{aligned}
$$

It is not hard to see that a derivation of the "same form" as the one just given could establish the absorption law for *set theory as well.* The following changes of symbolism in that derivation are all that are needed.

$\cup$ instead of $\vee$
$\cap$ instead of $\wedge$
$=$ instead of $\equiv$
$\emptyset$ instead of $\mathbf{F}$

Note that verifying the absorption law by the method of the preceding example yields more information than verifying it by truth tables. The method really shows that the absorption law holds not only in logic, but also in set theory, and, in fact, that it holds in any other "algebraic system" for which the commutative, identity, domination, and distributive laws hold. If a scientist of a future age were led by the needs of his science to devise such an "algebraic system," he would be sure (from our proof) that the absorption law holds in his system too.

This discussion illustrates what is often referred to as the **abstract approach.** This is the simultaneous study of a number of systems by first selecting or "abstracting" laws which are common to all of them and then deriving consequences of these laws.

The next section will continue in this spirit. Five laws which are common to logic and set theory (and other systems) will be selected. On the basis of these five laws, others will be derived. Algebraic systems satisfying the five laws selected are called **Boolean algebras** in honor of George Boole (1815–1864), who was one of the first people to use algebraic manipulations in the study of logic.

It would not be appropriate to use the symbolism of logic or the symbolism of set theory in the development of Boolean algebra since these are only special cases of Boolean algebras. Instead new symbols, whose meanings are left unspecified, are used. It is not important to know which operations and constants these symbols represent. It is only the laws which govern their behavior that count. Below are listed the new symbols, their names, and the way they can be interpreted in logic and set theory.

SYMBOL	NAME	LOGICAL INTERPRETATION	SET INTERPRETATION
$\cup$	cup	$\vee$	$\cup$
$\cap$	cap	$\wedge$	$\cap$
a'	complement of a	negation of a	complement of a
0	zero	F	$\emptyset$
1	one	T	U

Note: Be careful not to confuse 0 and 1 with the "numbers" 0 and 1 of ordinary arithmetic.

The word "complement" and its symbol were borrowed from set theory. This is unfortunate as the meaning is not always the same as in set theory. Nevertheless, this terminology is firmly rooted among practicing mathematicians, so you will have to live with it.

6.2
Definition of "Boolean Algebra"

If necessary, you should carefully reread the preceding section. Otherwise, the definition of the notion of a Boolean algebra will not seem to be well motivated.

Definition. An operation on a set S is a rule of correspondence which, to each pair of objects a, b of S (taken in that order), assigns exactly one object in S.

Example 1. Given:

R: the set of real numbers;
+: the usual addition of real numbers;

then "+" is an operation on R which assigns to each pair of real numbers, a and b, the number $a + b$.

Example 2. Given:

N: the set of *positive* whole numbers;
−: the usual subtraction of numbers;

then "−" is *not* an operation on N because even though a and b are in N, $a - b$ need not be a member of N. (It might be zero or a negative number.)

Definition. A **Boolean algebra** is a system which consists of a set $\mathcal{B}$ together with two operations $\cup$ and $\cap$, called **cup** and **cap** respectively, and which satisfies the following five laws:

B1. *Commutative laws:*

For all a, b $\in \mathcal{B}$,

$$a \cup b = b \cup a$$
$$a \cap b = b \cap a.$$

B2. *Associative laws:*

For all a, b, c $\in \mathcal{B}$,

$$a \cup (b \cup c) = (a \cup b) \cup c$$
$$a \cap (b \cap c) = (a \cap b) \cap c.$$

B3. *Distributive laws:*

For all a, b, c $\in \mathcal{B}$,

$$a \cup (b \cap c) = (a \cup b) \cap (a \cup c)$$
$$a \cap (b \cup c) = (a \cap b) \cup (a \cap c).$$

B4. *Identity laws:*

There exists in $\mathcal{B}$ objects (which we denote by 0, 1 respectively) with the property that

$$a \cup 0 = a \quad \text{for all } a \in \mathcal{B}$$

and

$$a \cap 1 = a \quad \text{for all } a \in \mathcal{B}$$

Objects with the property of 0 and 1 are called **identity** elements for the operations $\cup$ and $\cap$ respectively.

B5. *Complementation laws:*

For each a $\in \mathcal{B}$, there exists in $\mathcal{B}$ an object (which we denote by a′) with the property that

$$a \cup a' = 1$$
$$a \cap a' = 0.$$

Objects with the property of a′ are called **complements** of a.

In some books these laws are referred to as the **axioms** of Boolean algebra, while in others they are called the **postulates** of Boolean algebra.

Example 3. Given:

$\mathcal{B}$: the family of all subsets of a given universe U;
$\cup$: the operation of set union, $\bigcup$;
$\cap$: the operation of set intersection, $\bigcap$;

then $\mathcal{B}$ is a Boolean algebra with respect to the operations of $\cup$ and $\cap$. ϕ plays the role of the $\cup$-identity 0, while U plays the role of the $\cap$-identity 1.

Example 4. This example shows in precisely what sense the algebra of logic forms is a Boolean algebra. It is a very sophisticated example because the individual members of this Boolean algebra are themselves *sets* of logic forms.

Let $[\![\mathbf{A}]\!]$ denote the set of logic forms logically equivalent to the logic form **A**. Let $\mathcal{B}$ denote the family of all sets which are of the form $[\![\mathbf{A}]\!]$ for some **A**. The operations $\cup$ and $\cap$ on $\mathcal{B}$ are defined for $[\![\mathbf{A}]\!]$, $[\![\mathbf{B}]\!] \in \mathcal{B}$ as follows:

let

$$[\![\mathbf{A}]\!] \cup [\![\mathbf{B}]\!] = [\![\mathbf{A} \vee \mathbf{B}]\!],$$

and let

$$[\![\mathbf{A}]\!] \cap [\![\mathbf{B}]\!] = [\![\mathbf{A} \wedge \mathbf{B}]\!].$$

It is not hard to show that $\mathcal{B}$ is a Boolean algebra with respect to the operations $\cup$ and $\cap$ just defined.

$[\![\mathbf{F}]\!]$ plays the role of the $\cup$-identity 0.
$[\![\mathbf{T}]\!]$ plays the role of the $\cap$-identity 1.
$[\![\neg \mathbf{A}]\!]$ is the complement of $[\![\mathbf{A}]\!]$.

Remember: The members of $\mathcal{B}$ are not logic forms, but certain sets of logic forms. Two logic forms are in the same member of $\mathcal{B}$ if and only if they are logically equivalent.

Exercises 6.2

1. In this exercise you have the opportunity to verify that there is a Boolean algebra with exactly two members.

Let $\mathcal{B} = \{p, q\}$, where p and q are two distinct objects. Operations $\cup$ and $\cap$ on $\mathcal{B}$ are defined as follows:

$\cup$	$\cap$
$p \cup p = p$	$p \cap p = p$
$p \cup q = q$	$p \cap q = p$
$q \cup p = q$	$q \cap p = p$
$q \cup q = q$	$q \cap q = q$

Show that $\mathcal{B} = \{p, q\}$ is a Boolean algebra with respect to the operations of $\cup$ and $\cap$ just defined.

Which member of $\mathcal{B}$ plays the role of 0?
Which member of $\mathcal{B}$ plays the role of 1?
Which member of $\mathcal{B}$ is the complement of p?
Which member of $\mathcal{B}$ is the complement of q?

6.3
Duality

Suppose you have a statement in the symbolism of Boolean algebra and interchange the symbols in that statement as follows:

(a) interchange the symbols $\cup$ and $\cap$
(b) interchange the symbols 0 and 1

The resulting statement is called the **dual** of the original statement.

Example. The dual of $a \cup b = c$ is $a \cap b = c$.
The dual of $a \cup 0 = a$ is $a \cap 1 = a$.
The dual of $0 \cap 1 = 0$ is $1 \cup 0 = 1$.

If you carefully reexamine axioms **B1–B5** in the preceding section, you will discover the following curious and important fact:

The dual of each axiom is also an axiom.

This fact has the important consequence that if you prove a statement using the axioms of Boolean algebra, then you could also prove its dual. The reason for this is that you could replace each step in the proof by its dual. The resulting sequence of steps would be a proof of the dual of the original statement. This remarkable result is referred to as the **principle of duality.**

Example. Suppose you were to prove the idempotent law (this will be done later), $a \cup a = a$, from the axioms of Boolean algebra. Then the principal of duality guarantees that dual, $a \cap a = a$, would also be provable.

6.4 Some Theorems of Boolean Algebra

In this section some consequences of the axioms of Boolean algebra will be derived. *Remember:* Each result can be interpreted as a result of logic, or as a result of set theory.

Theorem I. In any Boolean algebra $\mathcal{B}$,

a) there is at most one element with the property of 0. That is, there is at most one identity element with respect to the operation $\cup$.
b) there is at most one element with the property of 1. That is, there is at most one identity element with respect to $\cap$.

PROOF: a) Suppose 0 and 0* were identity elements with respect to $\cup$; we shall show that $0 = 0^*$, i.e., 0 and 0* are the same element after all.

$0 = 0 \cup 0^* = 0^* \cup 0 = 0^*$
$\therefore 0 = 0^*$

b) The proof in this case is similar to the proof just given.

Theorem II. (IDEMPOTENT LAWS) In any Boolean algebra $\mathcal{B}$, for all $a \in \mathcal{B}$,

$$a \cup a = a$$
$$a \cap a = a$$

PROOF: By using the principle of duality, you need only show that one of the identities holds. Then the other also holds automatically. Now prove that $a \cup a = a$.

$a = a \cup 0$	**B4**	(identity law)
$= a \cup (a \cap a')$	**B5**	(complementation law)
$= (a \cup a) \cap (a \cup a')$	**B3**	(distributive law)
$= (a \cup a) \cap 1$	**B5**	(complementation law)
$= a \cup a$	**B4**	(identity law)

Theorem III. (DOMINATION LAWS) In any Boolean algebra $\mathcal{B}$, for all $a \in \mathcal{B}$,

$$a \cup 1 = 1$$
$$a \cap 0 = 0$$

PROOF: Because of the principle of duality, you have to verify only one of these identities to conclude that both hold. Prove that $a \cup 1 = 1$.

$1 = a \cup a'$	**B5**
$= a \cup (a' \cap 1)$	**B4**
$= (a \cup a') \cap (a \cup 1)$	**B3**
$= 1 \cap (a \cup 1)$	**B5**
$= (a \cup 1) \cap 1$	**B1**
$= a \cup 1$	**B4**

Theorem IV. (ABSORPTION LAWS) In any Boolean algebra $\mathcal{B}$, for all a, b in $\mathcal{B}$,

$$a \cap (a \cup b) = a$$

$$a \cup (a \cap b) = a$$

PROOF: Because of the principle of duality, you only have to verify one of these identities to conclude that both hold. Prove the top one.

$a \cap (a \cup b) = (a \cup b) \cap a$	**B1**
$= (a \cup b) \cap (a \cup 0)$	**B4**
$= a \cup (b \cap 0)$	**B3**
$= a \cup 0$	**Theorem III**
$= a$	**B4**

You may have noticed that this argument appeared before, in disguise. Go back to section 6.1 and take a look!

Theorem V. If in a Boolean algebra $\mathcal{B}$, there are elements a, b, and c such that both the relations

$$a \cup c = b \cup c$$

$$a \cap c = b \cap c$$

hold, then $a = b$.

PROOF:

$a = a \cup (a \cap c)$	**Theorem IV**
$= a \cup (b \cap c)$	**hypothesis**
$= (a \cup b) \cap (a \cup c)$	**B3**
$= (a \cup b) \cap (b \cup c)$	**hypothesis**
$= (b \cup a) \cap (b \cup c)$	**B1**
$= b \cup (a \cap c)$	**B3**
$= b \cup (b \cap c)$	**hypothesis**
$= b$	**Theorem IV**

Theorem VI. If in a Boolean algebra $\mathcal{B}$, there are elements a, b, and c such that both the relations

$$a \cup c = b \cup c$$

$$a \cup c' = b \cup c'$$

hold, then $a = b$.

PROOF:

$$\begin{aligned} a &= a \cup 0 && \textbf{B4} \\ &= a \cup (c \cap c') && \textbf{B5} \\ &= (a \cup c) \cap (a \cup c') && \textbf{B3} \\ &= (b \cup c) \cap (a \cup c') && \textbf{hypothesis} \\ &= (b \cup c) \cap (b \cup c') && \textbf{hypothesis} \\ &= b \cup (c \cap c') && \textbf{B3} \\ &= b \cup o && \textbf{B5} \\ &= b && \textbf{B4} \end{aligned}$$

Theorem VII. (UNIQUENESS OF COMPLEMENT) Each element a of a Boolean algebra $\mathcal{B}$ has at most one complement.

PROOF: Suppose a' and a^* were complements for a. Then, by the definition of complement,

$$\begin{matrix} a \cup a' = 1 & \text{and} & a \cup a^* = 1 \\ a \cap a' = 0 & & a \cap a^* = 0 \end{matrix}$$

thus:

$$a \cup a' = a \cup a^*$$

and

$$a \cap a' = a \cap a^*$$

therefore, by Theorem V, $a' = a^*$

Theorem VIII. In any Boolean algebra $\mathcal{B}$, for any $a \in \mathcal{B}$,

$$(a')' = a$$

PROOF: $a \cup a' = 1$ and $a \cap a' = 0$ by **B5.** Therefore, $a' \cup a = 1$ and $a' \cap a = 0$ by **B1.** Now this last pair of identities means that a is a complement for a′. But $(a')'$ is also a complement for a′. In fact, that is just what the symbolism $(a')'$ denotes. In Theorem VII it was proved that each element in $\mathcal{B}$ can have only one complement. Therefore, a and $(a')'$ must be the same object that is, $a = (a')'$.

Exercises 6.4

1. Prove de Morgan's laws. That is, show that in any Boolean algebra $\mathcal{B}$, for all $a, b \in \mathcal{B}$

$$(a \cup b)' = a' \cap b' \quad \text{and}$$

$$(a \cap b)' = a' \cup b'$$

[Hints: 1. Because of the principle of duality, only the top identity has to be verified. 2. Show that $a' \cap b'$ has the requisite properties to be a complement for $a \cup b$.]

2. Show that in a Boolean algebra $\mathcal{B}$, it can happen that $a \cap b = a \cap c$ even though $b \neq c$.

3. Show that the associative laws can be derived from the remaining four laws for a Boolean algebra.

[Hint: Let $x = a \cup (b \cup c)$
$y = (a \cup b) \cup c$.
Then show
$a \cap x = a \cap y$
$a' \cap x = a' \cap y$.
Then use the dual of Theorem VI to conclude that $x = y$.]

This result was first discovered by E. V. Huntington (1904). It shows that the associative laws could have been omitted from the definition of Boolean algebra.

7

Sample Examination

Time: One Hour

1. Make a truth table for the following logic form:

$$(A \wedge B) \vee (C \rightarrow B)$$

2. Find a logic form having the following for its table:

A	B	C	?
t	t	t	f
t	t	f	f
t	f	t	f
t	f	f	t
f	t	t	f
f	t	f	t
f	f	t	f
f	f	f	f

3. State in symbols:

   (a) Identity Laws (for logic)
   (b) Domination Laws (for logic)
   (c) Distributive Laws (for logic)
   (d) Absorption Laws (for logic)

4. Simplify:

$$[(A \wedge B) \vee (A \wedge B \wedge C)] \vee \neg A$$

5. Test for tautology (without using truth values):

$$[(A \wedge B) \rightarrow A] \wedge A$$

6. Test for contradiction (without using truth values):

$$(A \wedge \neg B) \rightarrow B$$

7. Simplify:

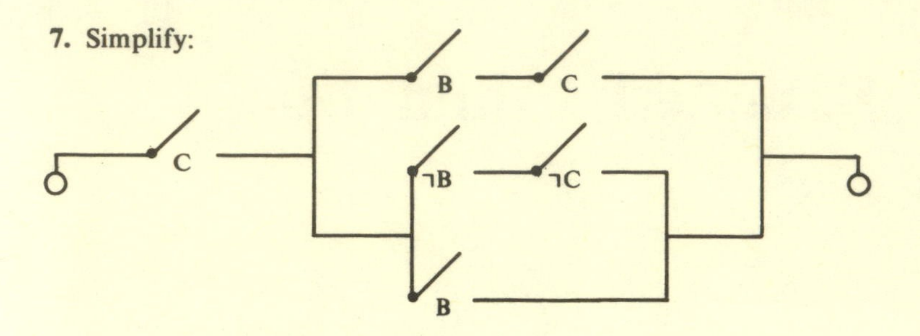

8. Write a logic form equivalent to

$$(A \wedge B) \rightarrow C,$$

but which uses only the operation symbols $\neg$, $\vee$.

9. Write a logic form equivalent to

$$A \rightarrow \neg B$$

but which has the Sheffer stroke as its only operation symbol.

10. Symbolize the following argument and check it for sentential validity.

> "Our mayor would keep his promises only if he were a man of integrity. But our mayor doesn't do this. So he lacks integrity."

11. In Cheerville, Illinois, all the 200 male inhabitants are devoted to wine, women, or song.

A poll showed:
100 are devoted to women
100 are devoted to wine
100 are devoted to song
20 are devoted to women and song
50 are devoted to women and wine
50 are devoted to wine and song

(a) How many are devoted to wine, women *and* song?
(b) How many are devoted to wine and song but not to women?

Answers

EXERCISES

SECTION 1.2

1. (a) It is not the case that geeks are foobles. ("Geeks are not foobles" is also an acceptable answer.)
   (b) Geeks are foobles and dobbies are tootles.
   (c) Geeks are foobles and it is not the case that dobbies are tootles.
   (d) If it is not the case that geeks are foobles, then dobbies are tootles.
   (e) It is not the case that if geeks are foobles, then dobbies are tootles.
   (f) Geeks are foobles if and only if dobbies are tootles.
   (g) If it is not the case that geeks are foobles or if dobbies are tootles, then geeks are foobles.
   (h) Geeks are foobles and dobbies are tootles if and only if it is not the case that dobbies are tootles.

2. (a) $A \wedge B$
   (b) $A \leftrightarrow B$
   (c) $\neg A \rightarrow B$
   (d) $A \rightarrow (B \vee C)$
   (e) $(B \wedge C) \rightarrow A$
   (f) $\neg B \leftrightarrow A$
   (g) $\neg (A \leftrightarrow B)$

3. (a) t
   (b) t
   (c) f
   (d) f
   (e) t
   (f) t
   (g) t
   (h) t
   (i) t
   (j) t
   (k) f
   (l) f
   (m) t
   (n) f

4. (a) t or f
   (b) no truth value
   (c) t
   (d) f
   (e) t or f

SECTION 1.3

1. (a) $B \wedge A$
   (b) $\neg D \wedge C$
   (c) $C \leftrightarrow \neg D$
   (d) $A \vee B$
   (e) $C \rightarrow D$

2. (a) $M \wedge J$
   (b) $(M \wedge J) \wedge T$
   (c) $(M \vee J) \rightarrow T$
   (d) $M \rightarrow J$
   (e) $\neg M \vee J$
   (f) $J \rightarrow M$
   (g) $T \rightarrow \neg J$

SECTION 1.4

1. The sentences written symbolically are:
   (a) $D \rightarrow R$
   (b) $T \rightarrow \neg R$
   (c) $R \rightarrow D$

   Their converses (in English) are:
   (a) If Ollie roars, then Ollie is a dragon.
   (b) If Ollie does not roar, then Ollie is toothless.
   (c) Ollie is a dragon only if Ollie roars.

   Their converses, written symbolically, are:
   (a) $R \rightarrow D$
   (b) $\neg R \rightarrow T$
   (c) $D \rightarrow R$

Their contrapositives (in English) are:

(a) If Ollie does not roar, then Ollie is not a dragon.

(b) If Ollie roars, then Ollie is not toothless.

(c) If Ollie is not a dragon, then Ollie does not roar.

Their contrapositives, written symbolically, are:

(a) $\neg R \rightarrow \neg D$

(b) $R \rightarrow \neg T$

(c) $\neg D \rightarrow \neg R$

2. (a) t

(b) t

(c) t or f

3. (a) A possible example is: If $2 + 2 = 4$, then $2 \cdot 2 = 4$. Of course, you could concoct many correct answers.

(c) A possible example is: If $2 + 2 = 5$, then $2 \cdot 2 = 4$. Many correct examples can be constructed.

SECTION 1.5

1. (a)

A	$A \rightarrow \neg A$
t	f
f	t

(b)

A	B	C	$(C \wedge \neg B) \vee (A \rightarrow B)$
t	t	t	t
t	t	f	t
t	f	t	t
t	f	f	f
f	t	t	t
f	t	f	t
f	f	t	t
f	f	f	t

(c)

A	C	$\neg(A \vee C)$
t	t	f
t	f	f
f	t	f
f	f	t

(d)

A	B	C	D	$[(A \to B) \vee (C \to \neg D)] \to (A \wedge \neg D)$
t	t	t	t	f
t	t	t	f	t
t	t	f	t	f
t	t	f	f	t
t	f	t	t	t
t	f	t	f	t
t	f	f	t	f
t	f	f	f	t
f	t	t	t	f
f	t	t	f	f
f	t	f	t	f
f	t	f	f	f
f	f	t	t	f
f	f	t	f	f
f	f	f	t	f
f	f	f	f	f

(e)

A	B	C	$[(B \vee A) \wedge C] \leftrightarrow [(B \wedge C) \vee (A \wedge C)]$
t	t	t	t
t	t	f	t
t	f	t	t
t	f	f	t
f	t	t	t
f	t	f	t
f	f	t	t
f	f	f	t

2. (a) t
   (b) t
   (c) t
   (d) indeterminable

3. (a) Let S be: The stock's value rises.
   Let D be: A dividend is declared.
   Let M be: The stockholders will meet.
   Let B be: The board of directors summons the stockholders.
   Let C be: The chairman of the board resigns.

   Then the logic form corresponding to the sentence is:

$$(S \vee D) \rightarrow [M \leftrightarrow (B \wedge \neg C)]$$

   (b) f; indeterminable since the truth value of M is unspecified.

4.

			Horace	Gladstone	Klunker
H	**G**	**K**	$\neg G \wedge K$	$\neg H \rightarrow \neg K$	$K \wedge (\neg H \vee \neg G)$
t	t	t	f	t	f
t	t	f	f	t	f
t	f	t	t	t	t
t	f	f	f	t	f
f	t	t	f	f	t
f	t	f	f	t	f
f	f	t	t	f	t
f	f	f	f	t	f

   (a) Horace and Klunker lied.
   (b) Horace and Klunker are innocent; Gladstone is guilty.
   (c) Horace and Klunker are guilty; Gladstone is innocent.

5. (a) Diplomat B is the spy.
   (b) Diplomat A is the spy.

6. Yes. The logic forms corresponding to the complied facts, in order, are:

A: $F \rightarrow [[(G \wedge \neg (L \vee R)] \vee [L \wedge \neg (G \vee R)]]$
$\vee [R \wedge \neg (L \vee G)]$

B: $(\neg G \wedge \neg L) \rightarrow F$

C: $\neg R \rightarrow [(F \wedge L) \vee (F \wedge G) \vee (G \wedge L)]$
D: $(L \wedge G) \rightarrow \neg R$.

Then the truth table discussed in the hint is:

F	G	L	R	A	B	C	D	
t	t	t	t	f	t	t	f	
t	t	t	f	f	t	f	t	
t	t	f	t	f	t	t	f	
t	t	f	f	t	t	f	t	
t	f	t	t	f	t	t	t	
t	f	t	f	t	t	f	t	
t	f	f	t	t	t	t	t	✓
t	f	f	f	f	t	t	t	
f	t	t	t	t	t	t	f	
f	t	t	f	t	t	t	t	✓
f	t	f	t	t	t	t	t	✓
f	t	f	f	t	t	f	t	
f	f	t	t	t	t	t	t	✓
f	f	t	f	t	t	f	t	
f	f	f	t	t	f	t	t	
f	f	f	f	t	f	f	t	

By examining the checked rows, you should be able to see that the registrar followed his instructions.

SECTION 1.6

1. (a)

A	$A \rightarrow A$
t	t
f	t

tautology

(b)

A	B	$A \rightarrow (B \rightarrow A)$
t	t	t
t	f	t
f	t	t
f	f	t

tautology

(c) contingency

A	B	C	$(A \vee B) \leftrightarrow (A \wedge \neg C)$
t	t	t	f
t	t	f	t
t	f	t	f
t	f	f	t
f	t	t	f
f	t	f	f
f	f	t	t
f	f	f	t

(d) contingency

A	C	$A \vee (\neg A \wedge C)$
t	t	t
t	f	t
f	t	t
f	f	f

(e) contingency

A	B	$A \rightarrow (\neg A \wedge B)$
t	t	f
t	f	f
f	t	t
f	f	t

(f) tautology

A	B	$\neg(A \wedge B) \leftrightarrow (\neg A \vee \neg B)$
t	t	t
t	f	t
f	t	t
f	f	t

SECTION 1.7

1. Form the conjunction

$$\neg (A \wedge B) \wedge \neg (A \wedge \neg B) \wedge \neg (\neg A \wedge B)$$
$$\wedge \neg (\neg A \wedge \neg B)$$

and examine its truth table.

A	B	$\neg(A \wedge B) \wedge \neg(A \wedge \neg B) \wedge \neg(\neg A \wedge B) \wedge \neg(\neg A \wedge \neg B)$
t	t	f
t	f	f
f	t	f
f	f	f

Hence the given logic forms are sententially inconsistent.

2. (a) Let A be: A recession will occur;
Let B be: Unemployment will decrease;
Let C be: Wage controls will be imposed.

The sentences given can then be symbolized as follows:

$$A \vee (\neg B \rightarrow C)$$

$$\neg A \rightarrow B$$

$$C \rightarrow \neg B$$

These sentences are sententially consistent.

(b) Let A be: Imports increase.
Let B be: Exports decrease.
Let C be: Tariffs are imposed.
Let D be: Devaluation occurs.

The sentences given can then be symbolized as follows:

$$(A \vee B) \rightarrow (C \vee D)$$

$$C \leftrightarrow (A \wedge \neg D)$$

$$B \rightarrow (\neg C \vee \neg A)$$

$$\neg D \vee (C \wedge B)$$

They are sententially consistent.

3. Their advice is inconsistent; they can't all be correct. When told about this, Fox muttered "A foolish consistency is the hobgoblin of small minds."

SECTION 1.8

1. (a) $(A \wedge B) \vee (A \wedge \neg B) \vee (\neg A \wedge B)$
(b) $(A \wedge \neg B) \vee (\neg A \wedge \neg B)$

(c) $(A \wedge \neg B \wedge C) \vee (A \wedge \neg B \wedge \neg C) \vee (\neg A \wedge B \wedge C)$
$\vee (\neg A \wedge \neg B \wedge \neg C)$

(d) $(A \wedge B \wedge C) \vee (A \wedge B \wedge \neg C) \vee (A \wedge \neg B \wedge C)$
$\vee (\neg A \wedge B \wedge C) \vee (\neg A \wedge B \wedge \neg C) \vee (\neg A \wedge \neg B \wedge C)$

2. $(A \wedge \neg A) \vee (B \wedge \neg B) \vee (C \wedge \neg C)$

3. The truth table for the unknown logic form should be:

A	B	Blanco's Answer	Unknown Logic Form
t	t	t	t
t	f	f	f
f	t	t	f
f	f	f	t

The table is constructed so that the value in Blanco's column agrees with the value in the B column and so that the logic form column agrees with Blanco's column only for those rows where A has value t. Therefore, the unknown logic form is

$$(A \wedge B) \vee (\neg A \wedge \neg B).$$

The question should be:

"Is it the case that either you tell the truth and the left-hand branch leads to the settlement, or you lie and the left-hand branch does not lead to the settlement."

SECTION 2.1

1. (a)

A	B	$(A \vee B) \leftrightarrow \neg(\neg A \wedge \neg B)$
t	t	t
t	f	t
f	t	t
f	f	t

$$A \vee B \equiv \neg(\neg A \wedge \neg B)$$

(b)

A	B	C	$(A \vee B) \leftrightarrow \neg(\neg B \wedge \neg C)$
t	t	t	t
t	t	f	t
t	f	t	t
t	f	f	f
f	t	t	t
f	t	f	t
f	f	t	f
f	f	f	t

$$A \vee B \not\equiv \neg(\neg B \wedge \neg C)$$

The symbol $\not\equiv$ means that the logic forms are not equivalent.

(c)

A	B	C	$A \wedge (B \vee C) \leftrightarrow (A \wedge B) \vee (A \wedge C)$
t	t	t	t
t	t	f	t
t	f	t	t
t	f	f	t
f	t	t	t
f	t	f	t
f	f	t	t
f	f	f	t

$$A \wedge (B \vee C) \equiv (A \wedge B) \vee (A \wedge C)$$

(d)

B	C	D	$C \wedge (B \vee D) \leftrightarrow (C \vee D) \wedge (B \vee D)$
t	t	t	t
t	t	f	t
t	f	t	f
t	f	f	t
f	t	t	t
f	t	f	t
f	f	t	f
f	f	f	t

$$C \wedge (B \vee D) \not\equiv (C \vee D) \wedge (B \vee D)$$

(e)

A	B	C	$A \to (B \to C) \leftrightarrow (A \to B) \to C$
t	t	t	t
t	t	f	t
t	f	t	t
t	f	f	t
f	t	t	t
f	t	f	f
f	f	t	t
f	f	f	f

$$A \to (B \to C) \not\equiv (A \to B) \to C$$

2.

A	B	C	$A \to (B \to C)$
t	t	t	t
t	t	f	f
t	f	t	t
t	f	f	t
f	t	t	t
f	t	f	t
f	f	t	t
f	f	f	t

The logic form:

$$(A \wedge B \wedge C) \vee (A \wedge \neg B \wedge C) \vee (A \wedge \neg B \wedge \neg C)$$
$$\vee (\neg A \wedge B \wedge C) \vee (\neg A \wedge B \wedge \neg C)$$
$$\vee (\neg A \wedge \neg B \wedge C) \vee (\neg A \wedge \neg B \wedge \neg C)$$

is logically equivalent to $A \to (B \to C)$ because it has the same truth table.

SECTION 2.3

1. (a) $A \vee B \equiv \neg\neg A \vee B$
$\equiv \neg A \to B$

(b) $\neg(A \rightarrow B) \equiv \neg(\neg A \vee B)$
$\equiv \neg\neg A \wedge \neg B$
$\equiv A \wedge \neg B$

(c) $A \rightarrow (B \rightarrow C) \equiv A \rightarrow (\neg B \vee C)$
$\equiv \neg A \vee (\neg B \vee C)$
$\equiv (\neg A \vee \neg B) \vee C$
$\equiv \neg (A \wedge B) \vee C$
$\equiv (A \wedge B) \rightarrow C$

(d) $(A \wedge B) \vee (\neg A \wedge C)$
$\equiv [A \vee (\neg A \wedge C)] \wedge [B \vee (\neg A \wedge C)]$
$\equiv (A \vee \neg A) \wedge (A \vee C) \wedge (B \vee \neg A) \wedge (B \vee C)$
$\equiv (A \vee C) \wedge (B \vee \neg A) \wedge (B \vee C)$
$\equiv (B \vee \neg A) \wedge (A \vee C) \wedge (B \vee C)$
$\equiv (\neg A \vee B) \wedge (A \vee C) \wedge (B \vee C)$
$\equiv (A \rightarrow B) \wedge (A \vee C) \wedge (B \vee C)$
$\equiv (A \rightarrow B) \wedge [(A \wedge B) \vee C]$
$\equiv (A \rightarrow B) \wedge [C \vee (A \wedge B)]$

2. (a) $\neg A \vee (A \vee B) \equiv (\neg A \vee A) \vee B$
$\equiv \neg A \vee A$
$\equiv \mathbf{T}$

(b) $\neg(\neg B \vee A) \equiv B \wedge \neg A$

(c) $\neg A \wedge (A \vee B) \equiv (\neg A \wedge A) \vee (\neg A \wedge B)$
$\equiv \neg A \wedge B$

(d) $(A \wedge B) \vee (A \wedge \neg B) \equiv A \wedge (B \vee \neg B)$
$\equiv A$

(e) $\neg(A \wedge \neg(A \vee B)) \equiv \neg A \vee \neg(\neg(A \vee B))$
$\equiv \neg A \vee (A \vee B)$
$\equiv (\neg A \vee A) \vee B$
$\equiv (\neg A \vee A)$
$\equiv \mathbf{T}$

(f) $A \wedge (A \vee B \vee C) \equiv (A \wedge A) \vee (A \wedge B) \vee (A \wedge C)$
$\equiv A \vee (A \wedge B) \vee (A \wedge C)$
$\equiv A \vee [A \wedge (B \vee C)]$
$\equiv A$

(g) $[(\neg B \vee A) \wedge A] \rightarrow A \equiv [A \wedge (\neg B \vee A)] \rightarrow A$
$\equiv [A \wedge (A \vee \neg B)] \rightarrow A$
$\equiv A \rightarrow A$
$\equiv \mathbf{T}$

(h) $(A \vee B) \wedge [(B \vee A) \wedge C] \equiv (A \vee B) \wedge [(A \vee B) \wedge C]$
$\equiv [(A \vee B) \wedge (A \vee B)] \wedge C$
$\equiv (A \vee B) \wedge C$

(i) $\neg [A \vee \neg (B \wedge C)] \equiv \neg A \wedge \neg [\neg (B \wedge C)]$
$\equiv \neg A \wedge (B \wedge C)$

(j) $A \wedge (\neg A \vee B) \equiv (A \wedge \neg A) \vee (A \wedge B)$
$\equiv A \wedge B$

(k) $A \vee [\neg A \vee (B \wedge \neg C)] \equiv (A \vee \neg A) \vee (B \wedge \neg C)$
$\equiv A \vee \neg A$
$\equiv \mathbf{T}$

(l) $[(\neg A \wedge \neg B) \vee \neg (A \vee B)] \rightarrow (\neg A \wedge \neg B)$
$\equiv [(\neg A \wedge \neg B) \vee (\neg A \wedge \neg B)] \rightarrow (\neg A \wedge \neg B)$
$\equiv (\neg A \wedge \neg B) \rightarrow (\neg A \wedge \neg B)$
$\equiv \neg (\neg A \wedge \neg B) \vee (\neg A \wedge \neg B)$
$\equiv \mathbf{T}$

(m) $[A \wedge (B \wedge C)] \vee [\neg A \wedge (B \wedge C)] \equiv (A \vee \neg A) \wedge (B \wedge C)$
$\equiv B \wedge C$

(n) First note that the dotted expression on the top line is a tautology.

$(E \wedge D) \wedge (\neg [\underset{\cdots\cdots\cdots\cdots\cdots\cdots\cdots\cdots}{(A \rightarrow B) \leftrightarrow (\neg B \rightarrow \neg A]} \rightarrow C)$

$\equiv (E \wedge D) \wedge (\neg [A \vee \neg A] \rightarrow C)$
$\equiv (E \wedge D) \wedge ((A \vee \neg A) \vee C)$
$\equiv (E \wedge D) \wedge (A \vee \neg A)$
$\equiv (E \wedge D)$.

3. (a) Either freedom of the press is not an important safeguard of liberty or, in protecting it, our courts have not played a major role.
   (b) Although a politician seeks the presidency and he has sufficient financial backing, he cannot afford to appear on nationwide television frequently.
   (c) A man has no self respect although he is contributing to a better society.
   (d) We can halt pollution although we do not act now.
   (e) Although a man cannot join the union, if he must relocate his family, he can find a job in that factory.
   (f) A worker can share in the company profits, but if he demands fewer fringe benefits, then he does not work harder.

[Remember, there are many correct answers to this question; we have just presented some possible answers.]

4. No, for example

$$A \wedge (A \wedge \neg A) \equiv A \wedge \neg A, \text{ but}$$

$$A \wedge \neg A \not\equiv \neg A$$

5. $\mathbf{A} \wedge (\mathbf{A} \vee \mathbf{B}) \equiv (\mathbf{A} \vee \mathbf{F}) \wedge (\mathbf{A} \vee \mathbf{B})$
$\equiv \mathbf{A} \vee (\mathbf{F} \wedge \mathbf{B})$
$\equiv \mathbf{A} \vee \mathbf{F}$
$\equiv \mathbf{A}$

6. $[T \to (A \vee E)] \wedge [\neg E \to \neg (W \wedge T)] \wedge [(A \wedge \neg E) \to \neg T]$
$\equiv [T \to (A \vee E)] \wedge [(A \wedge \neg E) \to \neg T] \wedge [(W \wedge T) \to E]$
$\equiv [\neg T \vee (A \vee E)] \wedge [\neg (A \wedge \neg E) \vee \neg T] \wedge [\neg (W \wedge T) \vee E]$
$\equiv [\neg T \vee A \vee E] \wedge [\neg A \vee E \vee \neg T] \wedge [\neg W \vee \neg T \vee E]$
$\equiv [\neg T \vee E \vee A] \wedge [\neg T \vee E \vee \neg A] \wedge [\neg W \vee \neg T \vee E]$
$\equiv [[\neg T \vee E \vee A] \wedge [\neg T \vee E \vee \neg A]] \wedge [\neg W \vee \neg T \vee E]$
$\equiv [[(\neg T \vee E) \vee A] \wedge [(\neg T \vee E) \vee \neg A]] \wedge [\neg W \vee (\neg T \vee E)]$
$\equiv [(\neg T \vee E) \vee (A \wedge \neg A)] \wedge [(\neg T \vee E) \vee W]$
$\equiv (\neg T \vee E) \vee [(A \wedge \neg A) \wedge W]$
$\equiv (\neg T \vee E) \vee [\mathbf{F} \wedge W]$
$\equiv (\neg T \vee E) \vee \mathbf{F}$
$\equiv (\neg T \vee E)$
$\equiv T \to E$

A person pays taxes only if he earned $1,000 in the past 12 months.

7. The given set of rules corresponds to the logic form

$$(W \to H) \wedge [(H \wedge F) \to W] \wedge (F \to \neg W).$$

Now simplify this logic form:

$(W \to H) \wedge [(H \wedge F) \to W] \wedge (F \to \neg W)$
$\equiv (\neg W \vee H) \wedge [\neg (H \wedge F) \vee W] \wedge (\neg F \vee \neg W)$
$\equiv (\neg W \vee H) \wedge [\neg H \vee \neg F \vee W] \wedge (\neg F \vee \neg W)$
$\equiv [(\neg W \vee H) \vee (F \wedge \neg F)] \wedge [\neg H \vee \neg F \vee W]$
$\wedge [(\neg F \vee \neg W) \vee (H \wedge \neg H)]$
$\equiv (\neg W \vee H \vee F) \wedge (\neg W \vee H \vee \neg F) \wedge (\neg H \vee \neg F \vee W)$
$\wedge (\neg F \vee \neg W \vee H) \wedge (\neg F \vee \neg W \vee \neg H)$

idempotent law

$\equiv (\neg W \vee H \vee F) \wedge (\neg W \vee H \vee \neg F) \wedge (\neg H \vee \neg F \vee W)$
$\wedge (\neg F \vee \neg W \vee \neg H)$

$\equiv (\neg W \vee H \vee F) \wedge (\neg W \vee H \vee \neg F) \wedge (\neg H \vee \neg F \vee W)$
$\wedge (\neg H \vee \neg F \vee \neg W)$
$\equiv [(\neg W \vee H) \wedge (F \wedge \neg F)] \wedge [(\neg H \vee \neg F) \wedge (W \wedge \neg W)]$
$\equiv (\neg W \vee H) \wedge (\neg H \vee \neg F)$
$\equiv (W \rightarrow H) \wedge (H \rightarrow \neg F)$

Thus the simplified rules are:

(a) The warriors shall be chosen from among the hunters.

(b) No hunter shall be a farmer.

Section 2.4

1. (a) is
   (b) is
   (c) is not
   (d) is
   (e) is
   (f) is not
   (g) is
   (h) is not
   (i) is not

Section 2.5

1. (a) $\neg A \rightarrow B \equiv \neg\neg A \vee B$
   $\equiv A \vee B$

   (b) $\neg(\neg B \vee A) \equiv B \wedge \neg A$

   (c) $(A \wedge B) \vee (A \wedge \neg B) \equiv (A + B)(A + B')$
   $\equiv AA + BA + AB' + BB'$

   (d) $\neg(A \wedge \neg(A \vee B)) \equiv \neg A \vee [\neg\neg(A \vee B)]$
   $\equiv \neg A \vee A \vee B$

   (e) $A \leftrightarrow B \equiv (A \rightarrow B) \wedge (B \rightarrow A)$
   $\equiv (\neg A \vee B) \wedge (\neg B \vee A)$

   (f) $A \wedge [B \vee (C \wedge \neg D)] \equiv A + [B(C + D')]$
   $\equiv A + BC + BD'$

   (g) $(A \rightarrow B) \wedge [C \vee (A \wedge B)] \equiv A'B + [C(A + B)]$
   $\equiv A'B + CA + CB$

   (h) $(A \wedge B) \vee [(A \wedge C) \vee (A \wedge \neg B)]$
   $\equiv (A + B)(A + C)(A + B')$
   $\equiv (A + B)(AA + CA + AB' + CB')$
   $\equiv AAA + BAA + ACA + BCA + AAB' + BAB'$
   $+ ACB' + BCB'$

SECTION 2.6

1. (a) tautology
   (b) toutology
   (c) tautology
   (d) not a tautology
   (e) tautology
   (f) not a tautology
   (g) not a tautology
   (h) not a tautology
   (i) not a tautology

SECTION 2.7

1. (a) is
   (b) is
   (c) is not
   (d) is
   (e) is
   (f) is not
   (g) is not
   (h) is

2. We work in ordinary notation; you may prefer to use arithmetical notation.

(a) $A \rightarrow B \equiv \neg A \vee B$

(b) $A \vee B$

(c) $\neg (A \vee B) \equiv \neg A \wedge \neg B$

(d) $A \leftrightarrow B \equiv (A \rightarrow B) \wedge (B \rightarrow A)$

$\equiv (\neg A \vee B) \wedge (\neg B \vee A)$

$\equiv [(\neg A \vee B) \wedge \neg B] \vee [(\neg A \vee B) \wedge A]$

$\equiv (\neg A \wedge \neg B) \vee (B \wedge \neg B) \vee (\neg A \wedge A) \vee (B \wedge A)$

$\equiv (\neg A \wedge \neg B) \vee (B \wedge A)$

(e) $A \wedge B \wedge C$

(f) $(A \vee B) \wedge (B \rightarrow C)$

$\equiv (A \vee B) \wedge (\neg B \vee C)$

$\equiv [(A \vee B) \wedge \neg B] \vee [(A \vee B) \wedge C]$

$\equiv (A \wedge \neg B) \vee (B \wedge \neg B) \vee (A \wedge C) \vee (B \wedge C)$

$\equiv (A \wedge \neg B) \vee (A \wedge C) \vee (B \wedge C)$

(g) $[\neg(B \rightarrow C)] \rightarrow A \equiv \neg[\neg(\neg B \vee C)] \vee A$
$\equiv \neg B \vee C \vee A$

(h) $\neg[(B \rightarrow C) \rightarrow A] \equiv \neg[\neg(\neg B \vee C) \vee A]$
$\equiv (\neg B \vee C) \wedge \neg A$
$\equiv (\neg B \wedge \neg A) \vee (C \wedge \neg A)$

(i) $(A \vee B) \wedge [(B \vee A) \wedge C] \equiv (A \vee B) \wedge (A \vee B) \wedge C$
$\equiv (A \vee B) \wedge C$
$\equiv (A \wedge C) \vee (B \wedge C)$

(j) $A \wedge [(\neg A \rightarrow B) \vee C] \equiv A \wedge [(A \vee B) \vee C]$
$\equiv (A \wedge A) \vee (A \wedge B) \vee (A \wedge C)$
$\equiv A \vee (A \wedge B) \vee (A \wedge C)$

(k) $(A \rightarrow B) \rightarrow [(B \rightarrow C) \rightarrow (A \rightarrow C)]$
$\equiv (\neg A \vee B) \rightarrow [(\neg B \vee C) \rightarrow (\neg A \vee C)]$
$\equiv \neg(\neg A \vee B) \vee [(\neg B \vee C) \rightarrow (\neg A \vee C)]$
$\equiv \neg(\neg A \vee B) \vee [\neg(\neg B \vee C) \vee (\neg A \vee C)]$
$\equiv (A \wedge \neg B) \vee (B \wedge \neg C) \vee \neg A \vee C$

SECTION 2.8

1. (a) not a contradiction, but is a contingency
(b) not a contradiction, but is a contingency
(c) contradiction
(d) not a contradiction, but is a contingency

2. (a) tautology
(b) tautology
(c) tautology
(d) contradiction

3. Let A be: The XYZ company builds a new factory
Let B be: Some of the XYZ company's machinery must be replaced
Let C be: The XYZ company can obtain a loan
Let D be: The XYZ company will be threated with bankruptcy

The sentences in the exercises correspond to the following logic forms:

$$(A \vee B) \rightarrow (C \vee D); \quad C \leftrightarrow (A \wedge \neg B)$$

$$B \rightarrow (\neg C \vee \neg A); \quad D \vee (B \wedge C).$$

Now take the conjunction of these and reduce it to a disjunctive normal form. You will see that it is not a contradiction, so the sentences are sententially consistent.

SECTION 2.9

1. (a) $[(A \to B) \to C] \wedge A \equiv \neg\,[\neg\,[(A \to B) \to C] \vee \neg\, A]$
$\equiv \neg\,[\neg\,[\neg\,(\neg\, A \vee B) \vee C] \vee \neg\, A]$

(b) $(A \wedge B) \vee (B \to \neg\, C) \equiv \neg\,(\neg\, A \vee \neg\, B) \vee (\neg\, B \vee \neg\, C)$

(c) $[A \vee (C \wedge D)] \to B \equiv \neg\,[A \vee (C \wedge D)] \vee B$
$\equiv \neg\,[A \vee \neg\,(\neg\, C \vee \neg\, D)] \vee B$

2. $A \vee B \equiv \neg\,(\neg\, A \wedge \neg\, B)$
$A \to B \equiv \neg A \vee B$
$\equiv \neg(\neg\neg A \wedge \neg\, B)$
$\equiv \neg\,(A \wedge \neg\, B)$
$A \leftrightarrow B \equiv (A \to B) \wedge (B \to A)$
$\equiv \neg\,(A \wedge \neg\, B) \wedge \neg\,(B \wedge \neg\, A)$

3. $(A \to B) \vee C \equiv \neg\,[\neg\,(A \to B) \wedge \neg\, C]$
$\equiv \neg\,[\neg\,(\neg\,(A \wedge \neg\, B)) \wedge \neg\, C]$
$\equiv \neg\,[A \wedge \neg\, B \wedge \neg\, C]$

4. $A \vee B \equiv \neg\, A \to B$
$A \wedge B \equiv \neg\,(\neg\, A \vee \neg\, B)$
$\equiv \neg\,(\neg\,\neg\, A \to \neg\, B)$
$\equiv \neg\,(A \to \neg\, B)$
$A \leftrightarrow B \equiv (A \to B) \wedge (B \to A)$
$\equiv \neg\,[(A \to B) \to \neg\,(B \to A)]$

5. $A \wedge (B \vee C) \equiv \neg\,[A \to \neg\,(B \vee C)]$
$\equiv \neg\,[A \to \neg\,(\neg\, B \to C)]$

6. (a) $A \wedge B \equiv \neg\,[\neg\,(A \wedge B)] \equiv \neg\,(A \mid B) \equiv (A \mid B) \mid (A \mid B)$

(b) $A \to B \equiv \neg\, A \vee B \equiv \neg\,(A \wedge \neg\, B) \equiv A \mid (\neg\, B)$
$\equiv A \mid (B \mid B)$

(c) $A \vee \neg\, B \equiv \neg\, B \vee A \equiv \neg\,(B \wedge \neg\, A) \equiv B \mid (\neg\, A)$
$\equiv B \mid (A \mid A)$

(d) $\neg\, B \to A \equiv (B \vee A) \equiv \neg\,(\neg\, B \wedge \neg\, A) \equiv \neg\, B \mid \neg\, A$
$\equiv (B \mid B) \mid (A \mid A)$

7. Stronger hint: Use an indirect proof. Suppose a contradiction could be built from $\to$ and $\vee$. Then there would be one of smallest length.

SECTION 3.1

1. (a) not sententially valid
(b) sententially valid
(c) sententially valid
(d) sententially valid
(e) sententially valid
(f) sententially valid
(g) sententially valid
(h) not sententially valid

2. (a) not sententially valid
(b) not sententially valid
(c) sententially valid
(d) not sententially valid
(e) sententially valid
(f) not sententially valid

3. Hint: Any inference of the form

$$\begin{array}{l} A \rightarrow B \\ \underline{B \qquad} \\ \therefore A \end{array}$$

in which A is a true statement will do.

4. Hint: Any inference of the form

$$\begin{array}{l} A \rightarrow B \\ \underline{A \qquad} \\ \therefore B \end{array}$$

in which B is a false statement will do.

5. Hint: Any inference of the form

$$\begin{array}{l} A \rightarrow C \\ \underline{B \rightarrow C \quad} \\ \therefore (A \vee B) \rightarrow C \end{array}$$

in which A and B are true statements while C is a false statement will do.

SECTION 3.2

1. I. Note that $[A \wedge (A \rightarrow B)] \rightarrow B$ is a tautology.
   II. Note that $[(A \rightarrow B) \wedge (B \rightarrow C)] \rightarrow (A \rightarrow C)$ is a tautology.
   III. Note that $[(A \rightarrow C) \wedge (B \rightarrow C)] \rightarrow [(A \vee B) \rightarrow C]$ is a tautology.

SECTION 3.3

1. (a) A
$A \rightarrow (A \vee \neg B)$ taut.
$A \vee \neg B$
$\neg B \vee A$
$B \rightarrow A$

(b) $\neg C \rightarrow \neg B$
$B \rightarrow C$
$A \rightarrow B$
$A \rightarrow C$
$C \rightarrow D$
$A \rightarrow D$

(c) C
$C \rightarrow \neg B$
$\neg B$
$A \rightarrow B$
$\neg B \rightarrow \neg A$
$\neg A$

(d) $(A \rightarrow B) \vee (A \rightarrow C)$
$(\neg A \vee B) \vee (\neg A \vee C)$
$\neg A \vee (B \vee C)$
$A \rightarrow (B \vee C)$
A
$B \vee C$

(e) $A \vee B$
$\neg A \rightarrow B$
$A \rightarrow B$
$(\neg A \vee A) \rightarrow B$
$\neg (\neg A \vee A) \vee B$
$(A \wedge \neg A) \vee B$
B

(f) $B \rightarrow A$
$\neg A \rightarrow \neg B$

$B \lor D$
$\neg B \rightarrow D$
$\neg A \rightarrow D$
$D \rightarrow C$
$\neg A \rightarrow C$
$A \lor C$

(g) $\neg A \lor \neg D$
$D \rightarrow \neg A$
D
$\neg A$
$\neg A \rightarrow (B \rightarrow \neg C)$
$B \rightarrow \neg C$
$\neg E \rightarrow B$
$\neg E \rightarrow \neg C$
$C \rightarrow E$

SECTION 4.1

1. (a) $(A \lor B) \land \neg A$
(b) $A \land (B \lor \neg B)$
(c) $(A \lor B \lor \neg C) \land (A \lor C)$
(d) $A \lor (\neg B \land \neg C) \lor (D \land \neg B) \lor (C \land A)$
(e) $A \land [(B \land \neg C) \lor (C \land (D \lor \neg A))]$

2. (a)

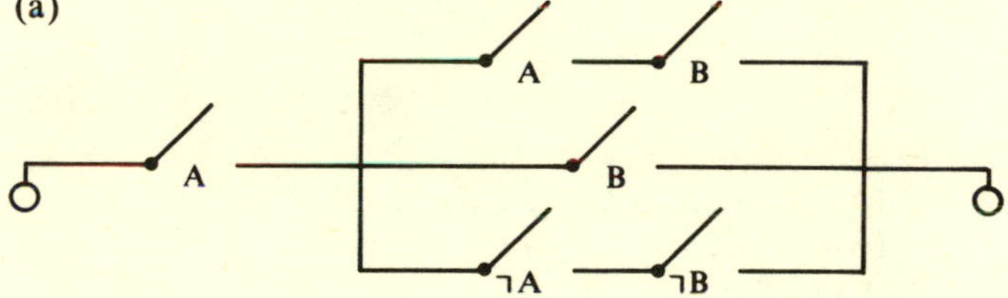

(b)

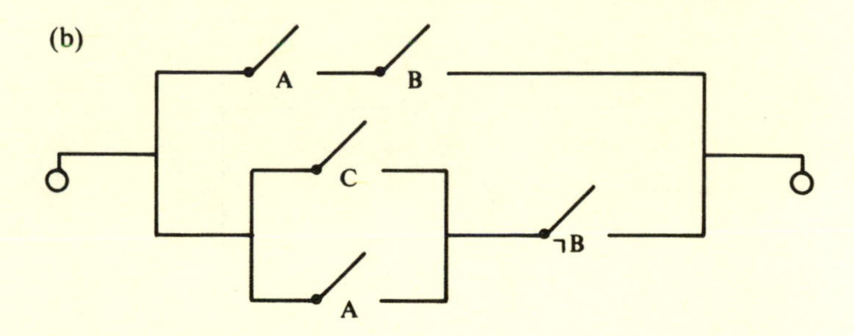

(c)

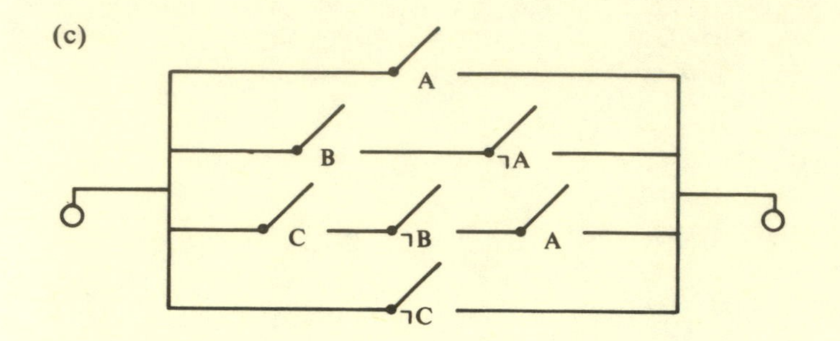

(d)

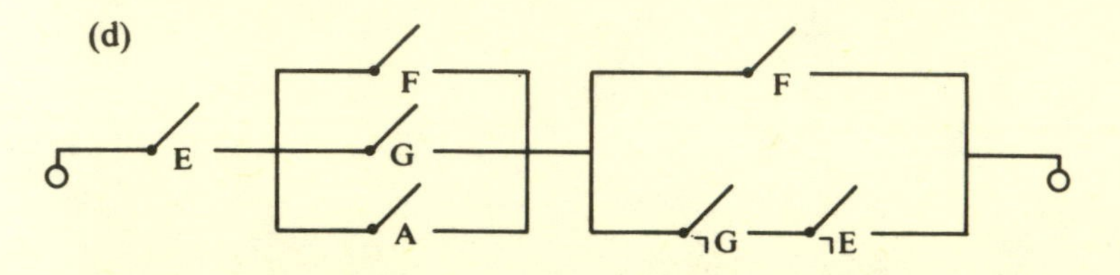

SECTION 4.2

1. (a) $A \wedge (\neg A \vee B) \equiv (A \wedge \neg A) \vee (A \wedge B)$
$\equiv A \wedge B$

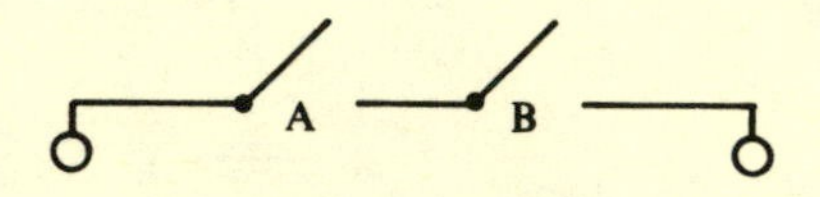

(b) $(A \wedge B) \vee [(C \vee A) \wedge \neg B)]$
$\equiv (A \wedge B) \vee (C \wedge \neg B) \vee (A \wedge \neg B)$
$\equiv (A \wedge B) \vee (A \wedge \neg B) \vee (C \wedge \neg B)$
$\equiv [(A \wedge (B \vee \neg B))] \vee (C \wedge \neg B)$
$\equiv A \vee (C \wedge \neg B)$

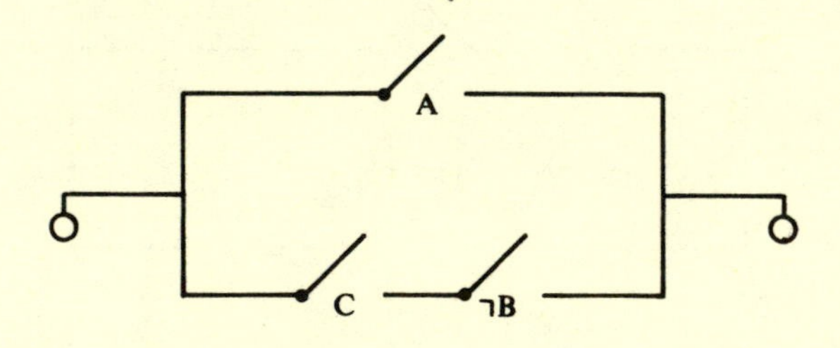

(c) $(B \wedge C) \vee (A \wedge \neg B \wedge C) \vee (\neg A \wedge \neg B \wedge C)$

$\equiv (B \wedge C) \vee [(A \vee \neg A) \wedge (\neg B \wedge C)]$

$\equiv (B \wedge C) \vee (\neg B \wedge C)$

$\equiv (B \vee \neg B) \wedge C$

$\equiv C$

(d) $(\neg A \vee \neg B) \wedge (\neg A \vee B) \wedge (A \vee B)$

$\equiv [\neg A \vee (B \wedge \neg B)] \wedge (A \vee B)$

$\equiv \neg A \wedge (A \vee B)$

$\equiv (\neg A \wedge A) \vee (\neg A \wedge B)$

$\equiv \neg A \wedge B$

(e) $A \wedge [(A \wedge B) \vee B \vee (\neg A \wedge \neg B)]$

$\equiv A \wedge [B \vee (\neg A \wedge \neg B)]$

$\equiv A \wedge [B \vee \neg A]$

$\equiv (A \wedge B) \vee (A \wedge \neg A)$

$\equiv A \wedge B$

(f) $(A \vee B \vee C \vee D) \wedge (A \vee B \vee D) \wedge (A \vee C)$

absorption

$\equiv (A \vee B \vee D) \wedge (A \vee C)$

$\equiv A \vee [(B \vee D) \wedge C]$

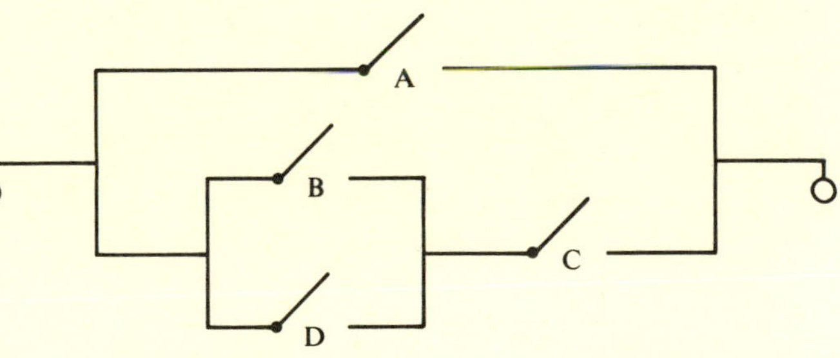

(g) $[C \wedge (A \vee \neg B)] \vee [\neg C \wedge A] \vee [(C \vee \neg B) \wedge \neg C]$
$\equiv [C \wedge (A \vee \neg B)] \vee [\neg C \wedge A] \vee [C \wedge \neg C] \vee [\neg B \wedge \neg C]$
$\equiv (C \wedge A) \vee (C \wedge \neg B) \vee (\neg C \wedge A) \vee (\neg C \wedge \neg B)$
$\equiv [(C \vee \neg C) \wedge A] \vee [(C \vee \neg C) \wedge \neg B]$
$\equiv A \vee \neg B$

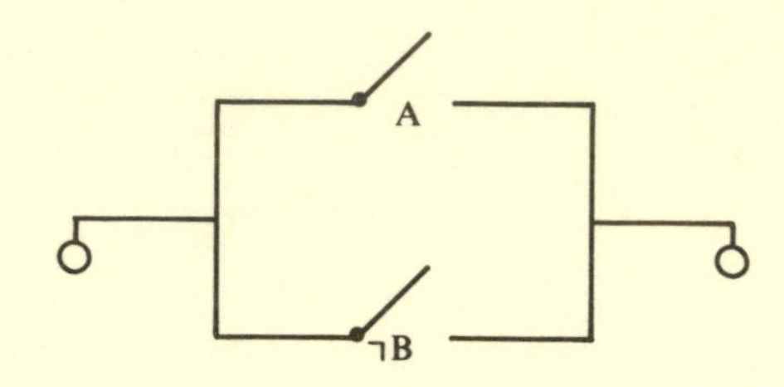

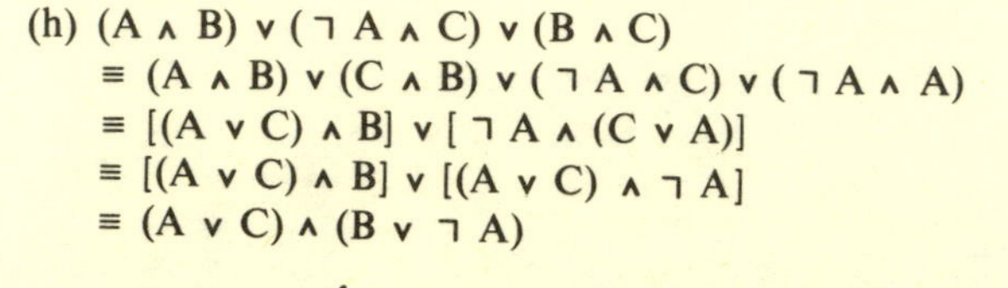

(h) $(A \wedge B) \vee (\neg A \wedge C) \vee (B \wedge C)$
$\equiv (A \wedge B) \vee (C \wedge B) \vee (\neg A \wedge C) \vee (\neg A \wedge A)$
$\equiv [(A \vee C) \wedge B] \vee [\neg A \wedge (C \vee A)]$
$\equiv [(A \vee C) \wedge B] \vee [(A \vee C) \wedge \neg A]$
$\equiv (A \vee C) \wedge (B \vee \neg A)$

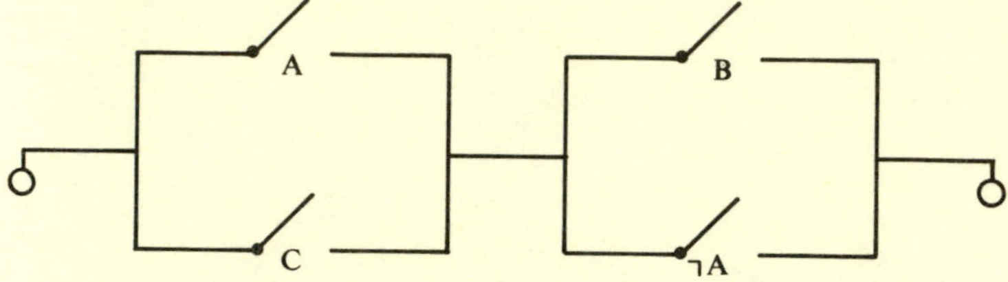

SECTION 4.3

1.

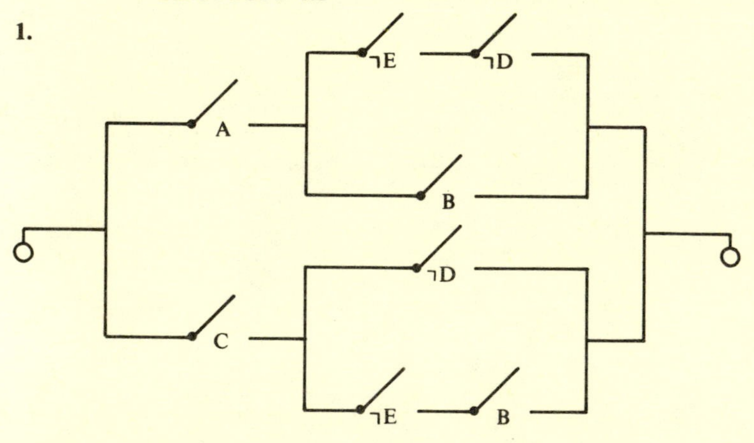

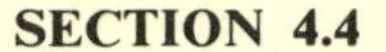

SECTION 4.4

1.

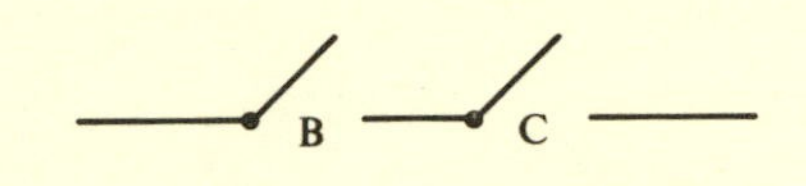

2. Stronger hint: Finish filling out the table below. To say that each switch can control the circuit is the same as saying that if, in two given rows, exactly two of the letters A, B, C get the same value, then the right-hand entries of those two given rows should be of opposite value. This information should enable you to complete the table.

A	B	C	??
on	on	on	on
on	on	off	
on	off	on	
on	off	off	
off	on	on	
off	on	off	
off	off	on	
off	off	off	

3. Stronger hint: Make a table letting A, B, C, D and E denote the switches controlled by each of the councilmen. If, in a given row, three or more of these letters get the value "on," then the right-hand entry for that row should get the value "on."

4. Here A denotes the button pushed by the prime minister, while

B, C, D, and E denote the buttons controlled by each of the other four advisors.

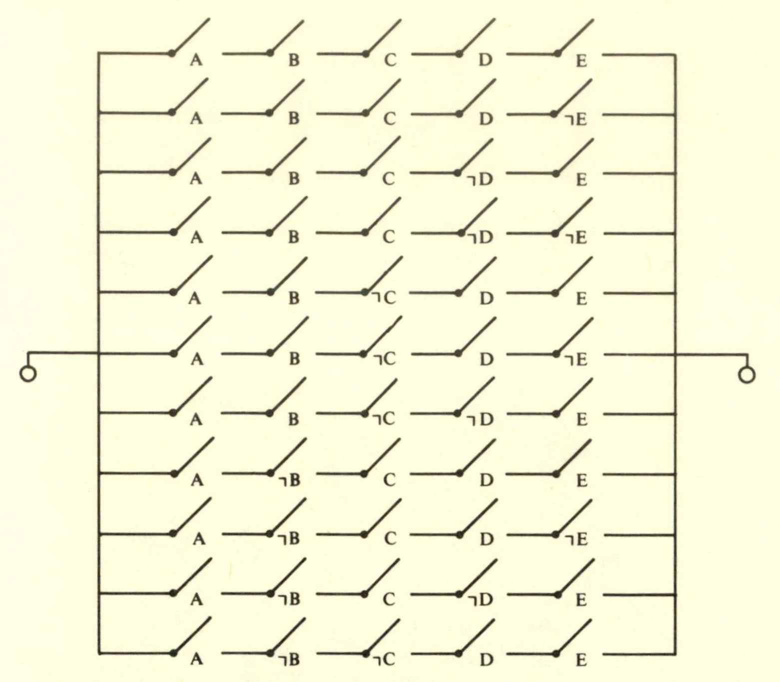

This circuit can, of course, be simplified.

SECTION 5.1

1. Yes, because each element of the first set is an element of the second set and vice versa.

2. (a) $\{d, e, x, y, z\}$

(b) $\{a, b, c, e, x\}$

(c) $\{a, b, c\}$

(d) $\emptyset$

(e) $\{a, b, c, d, e, x, y, z\}$

(f) $\{a, b, c, e, x, y, z\}$

(g) $\{a, e\}$

(h) $\{y, z\}$

3. $\{a\}$
$\{b\}$
$\{c\}$
$\{d\}$
$\{a, b\}$
$\{a, c\}$
$\{a, d\}$
$\{b, c\}$
$\{b, d\}$
$\{c, d\}$
$\{a, b, c\}$
$\{a, b, d\}$
$\{a, c, d\}$
$\{b, c, d\}$
$\emptyset$
$\{a, b, c, d\}$

4. (a) true
(b) true
(c) false
(d) true
(e) true
(f) true
(g) false
(h) false
(i) true
(j) true
(k) false
(l) false

SECTION 5.2

1. (a) $B \subseteq A'$ (or, equivalently, $A \subseteq B'$)
(b) $A = \emptyset$
(c) $A = B$
(d) $A = B$
(e) $A \subseteq B$
(f) $A \subseteq B'$ (or $B \subseteq A'$)
(g) always

2. (a) $x \in (A')' \equiv \neg(x \in A')$
$\equiv \neg(\neg(x \in A))$
$\equiv x \in A$

(b) $x \in (A \cap (B \cup C)) \equiv (x \in A) \wedge [x \in B \vee x \in C]$
$\equiv (x \in A \wedge x \in B)$
$\vee (x \in A \wedge x \in C)$
$\equiv (x \in A \cap B) \vee (x \in A \cap C)$
$\equiv x \in [(A \cap B) \cup (A \cap C)]$

(c) $x \in A \cap (A \cup B) \equiv (x \in A) \wedge (x \in A \vee x \in B)$
$\equiv x \in A$

(d) $x \in (A \cap B)' \equiv \neg(x \in (A \cap B))$
$\equiv \neg(x \in A \wedge x \in B)$
$\equiv \neg(x \in A) \vee \neg(x \in B)$
$\equiv x \in A' \vee x \in B'$
$\equiv x \in A' \cup B'$

(e) $x \in ((A \cup B) - (A \cap B))$
$\equiv x \in (A \cup B) \wedge \neg(x \in A \cap B)$
$\equiv (x \in A \vee x \in B) \wedge \neg(x \in A \wedge x \in B)$
$\equiv (x \in A \vee x \in B) \wedge [\neg(x \in A) \vee \neg(x \in B)]$
$\equiv [(x \in A \vee x \in B) \wedge \neg(x \in A)] \vee$
$[(x \in A \vee x \in B) \wedge \wedge \neg(x \in B)]$
$\equiv [x \in B \wedge \neg(x \in A] \vee [x \in A \wedge \neg(x \in B)]$
$\equiv x \in B - A \vee x \in A - B$
$\equiv x \in [(B - A) \cup (A - B)]$

(f) $x \in (A - B) \cup (A \cap B)$
$\equiv x \in A - B \vee x \in A \cap B$
$[x \in A \wedge \neg(x \in B)] \vee [x \in A \wedge x \in B]$
$\equiv (x \in A) \wedge [\neg(x \in B) \vee (x \in B)]$
$\equiv x \in A$

SECTION 5.3

1. Hint: By reassociating, the union of more than two sets can always be treated as just the union of *two* sets; e.g.,

$$A \cup B \cup C \cup D = A \cup [B \cup C \cup D]$$

Keep applying the formula for *two* sets

$$\#(X \cup Y) = \#(X) + \#(Y) - \#(X \cap Y)$$

over and over, using the distributive law where necessary. The answer finally is:

$$\begin{aligned}\#(A \cup B \cup C \cup D) = {} & \#(A) + \#(B) + \#(C) + \#(D) \\ & - \#(A \cap B) - \#(A \cap C) - \#(A \cap D) \\ & - \#(B \cap C) - \#(B \cap D) - \#(C \cap D) \\ & + \#(A \cap B \cap C) + \#(A \cap B \cap D) \\ & + \#(A \cap C \cap D) + \#(B \cap C \cap D) \\ & - \#(A \cap B \cap C \cap D)\end{aligned}$$

Closely examining the form of this answer, you should be able to make a good guess as to how the formula ought to look for calculating the number of elements in the union of *five* sets!

2. 35, 1, 7

3. 30%. Hint: Let M be the set of people losing their licenses, and let $m = \#(M)$. Then if

S is: the set of speeders,
R is: the set of people not stopping for a red light,
P is: the set of people passing in "no passing" zones,

$\#(S) = .6m, \quad \#(R) = .75m, \quad \#(P) = .70m,$
$\#(S \cap R) = .45m, \quad \#(S \cap P) = .50m, \quad \#(P \cap R) = .40m,$
and
$\#(M) = \#(S \cup P \cup R).$

4. (a) 22
(b) 14, 40, 2

SECTION 6.2

1. To show that $\mathcal{B}$ is a Boolean algebra, check that $\mathcal{B}$ satisfies axioms **B1–B5** by actually computing all the appropriate expressions.

"p" plays the role of 0
"q" plays the role of 1
$p' = q$
$q' = p$

SECTION 6.4

1. Because of the principle of duality, verify only

$$(a \cup b)' = a' \cap b'$$

(i) First note

$$\begin{aligned}(a \cup b) \cap (a' \cap b') &= [(a \cup b) \cap a'] \cap b' \\ &= [(a \cap a') \cup (b \cap a')] \cap b' \\ &= [0 \cup (b \cap a')] \cap b' \\ &= (b \cap a') \cap b' \\ &= (a' \cap b) \cap b' \\ &= a' \cap (b \cap b') \\ &= a' \cap 0 \\ &= 0\end{aligned}$$

(ii) Now note

$$\begin{aligned}(a \cup b) \cup (a' \cap b') &= a \cup [b \cup (a' \cap b')] \\ &= a \cup [(b \cup a') \cap (b \cup b')] \\ &= a \cup [(b \cup a') \cap 1] \\ &= a \cup (b \cup a') \\ &= a \cup (a' \cup b) \\ &= (a \cup a') \cup b \\ &= 1 \cup b \\ &= 1\end{aligned}$$

From (i) and (ii) you can see that $a' \cap b'$ has the requisite properties to be the complement of $a \cup b$. Hence, by Theorem VII, $(a \cup b)' = a' \cap b'$.

2. Let $\mathcal{B}$ be the Boolean algebra consisting of the subsets of a set X, with $\cup$ and $\cap$ being set union and intersection respectively. Let A and B be two distinct subsets of X. Then,

$$A \cup X = B \cup X = X, \text{ but } A \neq B.$$

3. Let $x = a \cup (b \cup c)$ and let $y = (a \cup b) \cup c$. Then,

$$\begin{aligned}a \cap x &= a \cap [a \cup (b \cup c)] \\ &= (a \cap a) \cup [a \cap (b \cup c)] \\ &= a \cup [a \cap (b \cup c)] \\ &= a\end{aligned}$$

$$
\begin{aligned}
a \cap y &= a \cap [(a \cup b) \cup c] \\
&= [a \cap (a \cup b)] \cup (a \cap c) \\
&= a \cup (a \cap c) \\
&= a
\end{aligned}
$$

Hence $a \cap x = a \cap y$

Moreover:

$$
\begin{aligned}
a' \cap x &= a' \cap [a \cup (b \cup c)] \\
&= (a' \cap a) \cup [a' \cap (b \cup c)] \\
&= 0 \cup [a' \cap (b \cup c)] \\
&= a' \cap (b \cup c)
\end{aligned}
$$

and

$$
\begin{aligned}
a' \cap y &= a' \cap [(a \cup b) \cup c] \\
&= [a' \cap (a \cup b)] \cup (a' \cap c) \\
&= [(a' \cap a) \cup (a' \cap b)] \cup (a' \cap c) \\
&= [0 \cup (a' \cap b)] \cup (a' \cap c) \\
&= (a' \cap b) \cup (a' \cap c) \\
&= a' \cap (b \cup c)
\end{aligned}
$$

Hence $a' \cap x = a' \cap y$. So by the dual of Theorem VI, $x = y$.

SAMPLE EXAMINATION

1.

A	B	C	$A \wedge B$	$C \rightarrow B$	$(A \wedge B) \vee (C \rightarrow B)$
t	t	t	t	t	t
t	t	f	t	t	t
t	f	t	f	f	f
t	f	f	f	t	t
f	t	t	f	t	t
f	t	f	f	t	t
f	f	t	f	f	f
f	f	f	f	t	t

Problems of this type are discussed in *Section 1.5.*

2. $(A \wedge \neg B \wedge \neg C) \vee (\neg A \wedge B \wedge \neg C)$

Problems of this type are discussed in *Section 1.8.*

3. (a) Identity Laws

$\mathbf{B \wedge T \equiv B}$
$\mathbf{B \vee F \equiv B}$

(b) Domination Laws

$\mathbf{B \vee T \equiv T}$
$\mathbf{B \wedge F \equiv F}$

(c) Distributive Laws

$\mathbf{A \wedge (B \vee C) \equiv (A \wedge B) \vee (A \wedge C)}$
$\mathbf{A \vee (B \wedge C) \equiv (A \vee B) \wedge (A \vee C)}$

(d) Absorption Laws

$\mathbf{A \wedge (A \vee B) \equiv A}$
$\mathbf{A \vee (A \wedge B) \equiv A}$

Consult *Section 2.2*.

4. Start with

$[(A \wedge B) \vee (A \wedge B \wedge C)] \vee \neg A$

Apply absorption law to the expression in the square brackets, treating $A \wedge B$ as a single quantity, to get

$[A \wedge B] \vee \neg A$

commute

$\neg A \vee [A \wedge B]$

distribute

$(\neg A \vee A) \wedge (\neg A \vee B)$

identity

$\neg A \vee B$

Consult *Section 2.3*.

5. Reduce to conjunctive normal form. Note that in the conjunctive normal form, at least one conjunct fails to contain a letter and also its negation. Thus, the logic form is not a tautology.

Consult *Section 2.6*.

6. Reduce to disjunctive normal form. Note that at least one disjunct fails to have a letter and also its negation. Thus, the logic form is not a contradiction.

Consult *Section 2.8*.

7. The logic form for this circuit is

$$C \wedge [(B \wedge C) \vee (\neg B \wedge \neg C) \vee B]$$

This can be manipulated into

$$B \wedge C.$$

Thus the simplified circuit is

B — C

Consult *Section 4.2*.

8. $(A \wedge B) \rightarrow C$

arrow law

$\neg (A \wedge B) \vee C$

de Morgan

$\neg A \vee \neg B \vee C$ Answer. (This, however, is not the only correct answer.)

Consult *Section 2.9*.

9. $A \rightarrow \neg B$

$\neg A \vee \neg B$

$\neg (A \wedge B)$

$A \mid B$ Answer. (This, however, is not the only correct answer.)

Consult *Section 2.9*.

10. Let P be: Our mayor keeps his promises.
Let I be: He is a man of integrity.

The argument in symbols is

$$\begin{array}{l} P \rightarrow I \\ \neg P \\ \hline \therefore \neg I \end{array}$$

Form the logic form

$$[(P \rightarrow I) \wedge \neg P] \rightarrow \neg I$$

This is not a tautology. Thus, the argument is not sententially valid.

Consult *Section 3.1*.

11. Let W be: Set of all those men who like women.
Let G be: Set of all those men who like wine (grapes).
Let S be: Set of all those men who like song.
Consulting *Section 5.3,* find the identity

$$\begin{aligned}\#(W \cup G \cup S) = {} & \#(W) + \#(G) + \#(S) \\ & - \#(W \cap S) - \#(W \cap G) - \#(G \cap S) \\ & + \#(W \cap G \cap S)\end{aligned}$$

Substituting given values, the result is

$$\begin{aligned}200 = {} & 100 + 100 + 100 - 20 - 50 - 50 \\ & + \#(W \cap G \cap S).\end{aligned}$$

Solving for $\#(W \cap G \cap S)$, you get

$$\#(W \cap G \cap S) = 20 \text{ Answer to part (a).}$$

To solve part (b) proceed as follows:

$$\begin{aligned}(G \cap S) & = (G \cap S) \cap (W \cup W') && \text{identity law} \\ & = (G \cap S \cap W) \cup (G \cap S \cap W') && \text{distributive law}\end{aligned}$$

Since the two sets which you are unioning on the right are disjoint, you can simply add up the number of members in each to get

$$\#(G \cap S) = \#(G \cap S \cap W) + \#(G \cap S \cap W').$$

Substituting the value given for $\#(G \cap S)$ and the value calculated in part (a) for $\#(G \cap S \cap W)$, you have

$$50 = 20 + \#(G \cap S \cap W')$$

$$\therefore \quad \#(G \cap S \cap W') = 30$$

Suggested Books for Further Reading

Elementary

Bowran, A. P. *A Boolean Algebra.* New York, Macmillan, 1965.

Dubisch, Roy. *Lattices to Logic.* New York, Blaisdell, 1963.

Hill, Shirley, and Suppes, Patrick. *First Course in Mathematical Logic.* New York, Blaisdell, 1964.

Quine, Willard Van Orman. *Elementary Logic.* New York, Blaisdell, 1964.

Tarski, Alfred. *Introduction to Logic.* New York, Oxford University Press, 1965.

Advanced

Hohn, F. E. *Applied Boolean Algebra.* New York, Macmillan, 1960.

Kneebone, G. F. *Mathematical Logic and the Foundations of Mathematics.* D. Van Nostrand, New York, 1963.

Lightstone, A. H. *The Axiomatic Method.* Prentice Hall, Englewood Cliffs, New York, 1964.

Mendelson, Elliott. *Boolean Algebra and Switching Circuits.* McGraw-Hill, New York, 1970.

Stoll, Robert R. *Set Theory and Logic.* W. H. Freeman, San Francisco, 1961.

Index